Reading and understanding applied statistics

A self-learning approach

Reading and understanding applied statistics

A self-learning approach

SIDNEY M. STAHL
Purdue University

JAMES D. HENNES
Colorado Department of Education

With 46 illustrations

The C. V. Mosby Company

Saint Louis 1975

Printed in the United States of America

International Standard Book Number 0-8016-4753-3

Distributed in Great Britain by Henry Kimpton, London

GW/VH/VH 9 8 7 6 5 4 3 2 1

To

Elaine

Heather
Juliet
Gabriel

Virginia

Teresa
Bryan
David
Mark

Preface

This book is designed to present basic statistical knowledge directly applicable in a variety of disciplines that share as a common element the study of humans and human behavior. It is aimed at students seeking their own introduction and understanding of applied statistics and at students or professionals who must be able to read critically the literature in their field for its statistical content.

The presentation of materials includes examples from the literature of sociology, psychology, education, and the health sciences. The material covers basic descriptive and inferential statistics as well as a brief introduction to scientific methodology. Also included is a brief chapter on the evaluation of scientific work from a methodological and statistical point of view. We have kept the formal mathematical materials to a minimum.

We feel strongly that the learning of statistics should be accomplished in a manner that the learner finds best suited to his/her needs. We have, therefore, prepared this introductory statistics book as a text that can be used either independently or in coordination with formal lecture materials. As the title suggests, this is a self-learning book, which, through its branched format, allows the reader to go rapidly over familiar or easily comprehensible materials. Specific learning objectives are provided for each chapter, and self-examinations are provided. The learner should be oriented toward mastering the objectives, not simply completing the book. To assure that these objectives were met, research was undertaken during the development of this program. Capable students, without any prior instruction in statistics, were able to master the objectives (excluding chapter 12 and part of chapter 7) at the 85% level with about 25 to 30 hours of study (Stahl and Hennes, 1973). The text has undergone several revisions based on extensive evaluation and the experiences gained in these earlier trials.

The book is organized into several conceptually different parts. The first five chapters represent what is classically called *descriptive statistics*. This is followed by a self-examination that tests the extent to which the learner has achieved the objectives. The examination's answers are keyed to specific objectives so that content material can be restudied. Chapters 6 through 10 present the basic issues in *inferential statistics* as well as the issues of methodology that are critical for an adequate understanding of inference. Chapter 11 covers the concept of association and ties it back into inference to complete the basic ideas in that broad area. Following chapter 11 is the second self-examination. The concluding chapter deals with further issues of methodology and suggestions on how to use the materials learned. It has been our experience that too frequently statistics books create a false division between methodology and statistics. The collection of data is dependent on some form of methodology. To understand the analysis of these data, since analysis is the primary role of statistics, it is important that the learner have some understanding of the principles of methodology. Therefore the few chapters that are included covering basic methodological issues are intended to bring these issues to the attention of the learner and to make more meaningful the statistics learned.

Appendix A consists of statistical tables used throughout the book. Appendix B provides answers for the two self-examinations. Appendix C presents formulas used throughout the book so that the learner will have a ready reference when needed. The reader is urged not to memorize the formulas, rather just to understand them.

The text is presented in *frames*. At the end of each frame follow the instructions of which frame to go to next. As you choose your answers, the instructions will often tell you to skip over explanation frames.

We would be remiss if we did not thank the many people who were significant in furthering our thinking on the issues of learning discussed in this preface and in the content of the self-learning book that follows. Although there were many people involved, some of the most helpful should be mentioned. To Hans O. Mauksch, who was instrumental in his inspiration and enthusiasm, a special debt of gratitude is due. To Merlyn Herrick, David Garloff, and the Education Resources Group in the School of Medicine, University of Missouri, special thanks are due for their technical assistance in the early work on this idea. Many of our colleagues were instrumental in getting some of our ideas in a form that would be presentable in a self-learning format; they include Barbara J. Bank, Ron Miller, Charles O. Carlton, Alan Pober, and Ben Ogbru. For help in finding sources, we are indebted to Mary Suelman Reed and Hope Waters. We are also grateful for the opportunity of serving on the American Statistical Association's Committee to Effect Optimal Medical-Statistical Interaction; it has provided a sounding board for many of our ideas. For the many long days and evenings spent typing the various phases of the manuscript we are especially grateful for the tolerance and help of Janet Bradley, Sheryl Russell, and Nancy Reed. Also helpful in this regard were Maria Sanchez, Nancy McDermott, Sue Simmons, Cindy Batta, and Cathy Bloodworth; thank you.

We are indebted to the Literary Executor of the late Sir Ronald A. Fisher, F.R.S., to Dr. Frank Yates, F.R.S., and to Longman Group, Ltd., London, for permission to reprint tables from their book *Statistical Tables for Biological, Agricultural and Medical Research.*

Perhaps most significant was the help and criticism of the many students in the School of Medicine at the University of Missouri and the Departments of Sociology at the University of Missouri and Purdue University. Without their input, we would have learned less about how people learn. Many of their ideas are incorporated throughout this book. Their concern and interest in learning and understanding has provided a continuing motivation for us.

We are especially grateful to our wives, Elaine and Ginny, who watched us agonize over the concepts of self-learning and the presentation of an idea we believed important. Their tolerance and enthusiasm have been very important to us.

Sidney M. Stahl
James D. Hennes

Origins of the self-learning text

In 1970, Hans O. Mauksch, Chief of the Section of Behavioral Sciences of the School of Medicine and Professor of Sociology of the University of Missouri, suggested that we begin thinking about how to "teach" statistics in an innovative manner to the medical students at the University of Missouri. We realized that the traditional classroom lecture approach was wasteful of time and that much more could be learned by the student if we concerned ourselves with using time, not content, as the learning variable. There had to be a way to excite students about a traditionally dull subject and at the same time allow them to proceed at their own pace.

To undertake this task, we established two long-range goals: (1) students must be able to perform those statistical procedures required in other course work while in college; (2) students must be able to read critically the research literature for its statistical content. Coupled with these goals is the philosophy that students must be allowed to learn at their own speed. The implementation of this philosophy allows for students to test their knowledge when they believe that they are prepared, not when the faculty is ready to give an examination. In this way, learning is for mastery, not for amount of material covered in a given unit of time.

Data gathered on the original program from first-year medical students indicated that all students, given the objectives, were eventually able to reach mastery (Stahl and Hennes, 1973). Regardless of the level of prior exposure to statistics, if any, the students scored, as a group, 85% on the mastery examination after a median of 25 to 30 hours of study. Study time included the program in its original form (excluding part of chapter 7 and all of chapter 12), peer discussions and, for some, reading other texts or consulting with faculty. Those without prior statistical instruction spent 10% to 25% more time reaching mastery than those with some previous exposure to statistics.

Since the original writing of the self-learning materials, a good deal of experimentation has gone into our revision. Originally the content material was written for a simple branching learning machine. Many students felt hindered by being tied to a machine in their learning center and being unable to have easy access to any chapter or section of a chapter they chose. We have experimented with rearranging various items and topical areas, with writing and rewriting test items, and with presenting the materials in different formats. In addition, several other schools and disciplines have experimented with the content material contained in this self-learning text and have provided valuable insight into the learning process that students go through in understanding statistical materials.

We have been impressed by the interest of our colleagues in the social sciences and health sciences, such as public health, nursing, medicine, and veterinary medicine, in our approach to self-learning in statistics. Based on this interest and their encouragement, we undertook to rewrite much of the material in our original work and to use examples from sociology, psychology, and education to complement the examples and material from medicine and nursing.

It is our conviction that the learning of statistics does not have to be an onerous chore. It is not necessary to approach all statistical learning in the usual classroom approach; this denies the variability of learning styles in any heterogeneous group of students. Furthermore, it has become painfully obvious that many "practitioners" of the disciplines at which this nontraditional book is aimed are either inadequately prepared or have forgotten much of their statistical background.

Thus this book is written with the expectation that it will facilitate the learning or relearning of basic statistical materials. The examples found throughout the book are, where possible, taken from the literature of sociology, psychology, education, medicine, and nursing. It is hoped that this feature will make the learning of statistics more immediately relevant for the learner.

Self-learning can take many formats. Based on our experience, we believed that keeping learners actively involved in the learning process would provide the best form of self-instruction. To accomplish this end, we chose a branched self-learning format. In this format learners are frequently presented with several alternatives; choice of an alternative will dictate the direction in which each one goes. At the conclusion of each set of frames, or each chapter, each learner will have covered identical content material aimed at meeting the objectives of that chapter.

In addition to the branched self-learning format, it is our conviction that providing the learner with objectives to be accomplished by the end of the chapter will facilitate his/her learning. Such conviction is supported by research in behavioral objectives (Duchastel and Merrill, 1973). Thus each chapter is preceded with objectives.

How to use this branched self-learning text

Since it is a basic tenet of this book that you should be able to make your own progress at the speed you feel to be most comfortable, it is necessary to bring several features of this book to your attention.

1. Each chapter is prefaced with specific learning objectives. At the conclusion of each chapter, you should be able to meet those objectives. In this way the learning objectives serve as a learning guide and should direct your reading to specific content materials in that chapter.

2. At the end of chapter 5 and again at the end of chapter 11, there is a self-grading examination. The examinations represent a selection of learning objectives that have been turned into questions. Appendix B provides answers to these questions. In addition, each answer is keyed to the specific learning objective and chapter on which it was based. If you miss the answer, you should go back to that objective and chapter and restudy the materials.

3. So that you will find immediate application of the principles presented in the book, we have tried to use and cite examples from the literature or to prepare examples that are similar to those frequently encountered. If you are interested in following the statistic into the literature, this should provide the beginning of that search. The citations are found throughout the text, and the references are listed at the end of the book. Any minor changes between the original data presented in the various journals and this book were made for didactic purposes and do not change the conclusions of the original work.

4. The book is organized in chapters and, within chapters, into *frames*. The frames are numbered in the upper left-hand corner and are in sequential order.

Read the entire frame, and then follow the instructions listed at the bottom of the frame. For example, if the frame is a question frame, you will be asked a question based on learned materials. The various possible answer choices will be listed as A, B, C, and D. Make a decision on which of the answers is correct and then go to the bottom right-hand corner of the frame and follow the instructions as to which frame to go to from there. For example, if you choose answer B, the statement may be "(B, go to 21)," which means that to determine if the correct answer was chosen, you must turn to frame 21. If your answer was incorrect, frame 21 will tell you why and attempt to present the materials in a different manner so that you can learn why you missed the question, as well as what the correct answer was. Because each question choice requires a separate answer frame, not every learner will read all the frames in the book. The various answer choices are based on a good deal of research and questioning of students as to what kinds of decisions were made. The choices then represent, essentially, a set of answers that you might logically make and, in fact, that other learners have made, based on varying degrees of understanding of a given question.

Contents

chapter
1

Some basic concepts in statistics

At the conclusion of this chapter, you should be able to:

1. Differentiate between descriptive and inductive statistics, for example, by comparing two statistical statements and identifying them as either inductive or descriptive.
2. Define the following: population, sample, data, statistics, variable.
3. Understand that valid statistical conclusions can be made only from valid data as, for example, being able to judge the accuracy of a conclusion from a statistical study.
4. Name the four major scales of measurement: nominal, ordinal, interval, ratio.
5. State which of the following characteristics are associated with each of the four scale types: mutually exclusive categories, exhaustive categories, ordered values, equal or quantifiable intervals between scale points, zero origin point in scale.
6. State which examples of scales satisfy the requirements of mutually exclusive and exhaustive categories.
7. State the scale of measurement for each scale on a list of scales or measurements.
8. State which of the following statements (or equivalents) is appropriate for each of the four scale types: value A is different from value B, value A is greater then value B, A is 10 points greater than B, $\frac{A + B}{2} = C$, A is twice as large as B.

1 Science operates on the basis of probabilistic knowledge. The functions of statistics are (1) to reduce large amounts of information into a more manageable size, (2) to provide an estimate or model of unknown values based on data at hand, (3) to determine to what extent a particular hypothesis or series of hypotheses can be supported, and (4) to provide an estimation of the risk involved in accepting these hypotheses as a basis for making a decision or taking action.

More directly the purpose of this statistics self-teaching approach is to:

1. Enable you to begin to evaluate, using scientific criteria, the adequacy of research designs and statistical analyses in the literature
2. Develop your own statistical skills and knowledge necessary to function as an informed professional, practitioner, or scholar
3. Develop an awareness of your own depth of knowledge
4. Introduce to you the skills available in the form of either future study or consultation with a professional statistician

The general study of statistics is usually divided into two topical areas: descriptive and inductive statistics. *Descriptive* statistics or, more accurately, the *descriptive statistical* method is any treatment of numerical data that does not involve making generalizations from a sample to a population. That is, when we are interested in describing a group of elements (people, test scores, etc.), we use descriptive statistics. When we make generalizations, predictions, estimations, or otherwise arrive at decisions in the face of uncertainty, we are using *inductive statistics* by a process called *statistical inference.*

(*Go to* 2)

---

2 Classify these examples as applications of descriptive or inductive statistics:

1. The average income of families who send their children to a particular day-care center is $7342.

2. There are 24 normal color-perception subjects and 6 color-blind subjects in a visual acuity experiment. There is a high probability that the first subject seen on a random assignment of subjects will have normal color perception.

A. Example 1—descriptive
Example 2—inductive
B. Example 1—inductive
Example 2—descriptive
C. Both examples are descriptive
D. Both examples are inductive

(A, *go to* 4) (C, *go to* 5)
(B, *go to* 6) (D, *go to* 3)

3 In example 1 an average income of $7342 is a statement of fact derived from a simple mathematical treatment of numerical data—the income of families with children in the day-care center. This was not an attempt to generalize to a population; it is an example of *descriptive* statistics.

In example 2, with 24 normal color-perception subjects and 6 color-blind subjects, a prediciton was made that the first subject seen in the experiment would be color-normal. However, it could also have been a color-blind subject. A color-normal subject is predicted on the basis of the distribution of all subjects in the experiment, but the prediction itself is an application of *inductive* statistics. This half of your answer was correct.

(*Go to* 7)

4 Your choice of example 1 as descriptive and example 2 as inductive is correct. An average is obtained by simple treatment of numerical data, which is a good example of *descriptive* statistics.

Similarly, the application of probability in predicting that with 24 normal color-perception subjects and 6 color-blind subjects in the experiment, the first subject seen on a randomly selected basis will be a normal color-perception subject, is a good example of statistical inference, i.e., *inductive* statistics.

(*Go to* 7)

5 In example 1 an average income of $7342 is a statement of fact derived from a simple mathematical treatment of numerical data—the income of families with children in the day-care center. This was not an attempt to generalize to a population; it is an example of *descriptive* statistics.

In example 2 with 24 normal color-perception subjects and 6 color-blind subjects, a prediction was made that the first subject seen in the experiment would be color-normal. However, it could also have been a color-blind subject. A color-normal subject is predicted on the basis of the distribution of all subjects in the experiment, but the prediction itself is an application of *inductive* statistics. Only the first half of your answer was correct.

(*Go to* 7)

6 In example 1 an average income of $7342 is a statement of fact derived from a simple mathematical treatment of numerical data—the income of families with children in the day-care center. This was not an attempt to generalize to a population; it is an example of *descriptive* statistics.

In example 2 with 24 normal color-perception subjects and 6 color-blind subjects, a prediction was made that the first subject seen in the experiment would be color-normal. However, it could also have been a color-blind subject. A color-normal subject is predicted on the basis of the distribution of all subjects in the experiment, but the prediction itself is an application of *inductive* statistics.

(*Go to* 7)

---

7 Here we will explore four terms important to the area of statistics:

1. Population. All people (or objects, or test scores, or . . .) in a specified area having at least one characteristic in common. For example, all test scores for students taking an examination, all diastolic blood pressure readings of patients in a given clinic, and all religious affiliations of persons in a given social group.

2. Sample. A subset of the population. Usually we take great pains to be certain that the sample of a population is representative of the population (i.e., randomly selected); a truly representative sample is said to be an *unbiased* sample. For example, a randomly selected subset of white mice from a popualtion of "maze-smart" mice is a sample from a population. Of course, samples also can be biased, nonrandom, and unrepresentative. Unfortunately, the literature contains many conclusions based on information from biased samples.

3. Data. Numbers or measurements that are collected as a result of observations of a sample (or population). The test scores of students taking an examination in a given course are an example of data.

4. Statistic. A number resulting from the manipulation of raw data collected from a sample, according to specified procedures. For example, the percentage of patients in a metabolic clinic whose blood cholesterol level is above the tolerable limit for their age is a statistic.

(*Go to* 8)

---

8 In a small group situation, one respondent felt that he was highly alienated from the remainder of the group, and another thought that he did not want to be a group member. This information is (are):

A. A population
B. A sample
C. Data
D. Statistics

(A, *go to* 10) (C, *go to* 11)
(B, *go to* 9) (D, *go to* 12)

---

9 The small group situation is not a sample; you will need to go back and review these concepts. Your answer is not correct.

(*Go to* 7)

---

10 You will need to go back and review these concepts; your answer (a population) is not correct.

(*Go to* 7)

---

11 The descriptive information about respondent feelings in the small group constitute data. Your answer is correct.

(*Go to* 13)

---

12 There is no evidence of data manipulation in this example; therefore these cannot be statistics. You will need to go back and review these concepts.

(*Go to* 7)

---

13 Which term best applies to the following?

The set of all children examined for learning disability problems by the psychology department clinic of a major state university.

A. Population
B. Sample
C. Data
D. Statistic

(A, *go to* 17) (C, *go to* 15)
(B, *go to* 14) (D, *go to* 16)

---

14 To use the techniques of statistical inference, it is necessary that a sample be unbiased and representative of the population about which statements are to be made. It is highly unlikely that these children with learning disabilities would be considered a sample. Since the children seen by the psychology department clinic do not systematically represent all children with learning disabilities in the community or any other known group, these children should be treated as a population.

(*Go to* 18)

---

15 Recall that data consist of numbers, measurements, or some form of observation. The children with learning disabilities examined at the psychology department clinic would best be described as a population.

(*Go to* 18)

---

16 Recall that a statistic results from some manipulation of raw data from a sample. The children examined for learning disabilities in the psychology department clinic would best be described as a population.

(*Go to* 18)

---

17 Since the children being examined for learning disabilities at the psychology department clinic do not systematically represent all children with learning disabilities in the community or any other known group, these children should be treated as a population.

(*Go to* 18)

---

18 Statistics, like most useful tools, can be misused. Experts in statistics may misuse statistics in an attempt to serve their special interests. Those who are not statisticians may misuse statistics by drawing conclusions or making claims that the statistics do not actually support. Statistical tests can be no more valid than the data being tested. The application of statistical methods to invalid data will assuredly produce results, but the results will be as false as the data from which the statistics were computed. Computer experts have a saying for this phenomenon: GARBAGE IN = GARBAGE OUT.

Sir Josiah Stamp, a British statistician, also had some thoughts in this regard: "The government are very keen on amassing statistics. They collect them, raise them to the *nth* power, take the cube root, and prepare wonderful diagrams. However, you must never forget that every one of these figures comes in the first instance from the village watchman, who just puts down what he damn pleases."

(*Go to* 19)

---

19 An elementary school textbook salesman reported that a certain book had proved to be statistically significantly more useful in retention of knowledge for children. One thousand school children in town X had used the text that he was selling, whereas 1000 school children in town Y used a competing text. The children in town X had 30% greater retention after a year than did the children of town Y. An examination of the test situation revealed that the children were closely matched in age, general health, and competence of teaching staff. Town X is a college community; town Y is a coal mining community. What do you think of this "research"?

A. Conclusion is probably accurate
B. Conclusion is probably inaccurate
C. I need more information

(A, *go to* 20)
(B, *go to* 21)
(C, *go to* 22)

---

20 The conclusion drawn about the statistical significance of this experiment is probably not accurate because a major variable was overlooked—the levels of education in the two towns. There is ample evidence that the educational backgrounds of families play important roles in the learning capabilities of children. Retention of materials would be expected to be greater in the college community than in the coal mining community. However, if we had information that parental educational background enhanced knowledge retention by a factor well under 30%, we would then have reason to believe the salesman's statistics. With the information given, his statistics are at least suspect, and the conclusion is inaccurate.

(*Go to* 23)

---

21 The conclusion is inaccurate; there is ample evidence that would suggest that the educatonal backgrounds of families play important roles in the learning capabilities of children. Retention of materials would be expected to be greater in a college community than in a coal mining community. The salesman is inaccurate in interpreting his data as favoring his product *unless* statistics can be produced showing that, in general, parental educational background enhances knowledge retention by a factor significantly lower than the 30% determined in this experiment. We really do not have valid data on which to base the salesman's conclusion.

(*Go to* 23)

---

22 You indicated that you need more information. Let us reexamine the data:

1. One thousand children in town X used the textbook in question.
2. One thousand children in town Y used a competing text.
3. After a year, the children in town X had 30% greater retention of learned materials.
4. The two groups of children were essentially matched except that town X is a college community, whereas town Y is a coal mining community.

There is ample evidence to suggest that the educational backgrounds of families play important roles in the learning capabilities of children. Thus we do not know whether the 30% greater knowledge retention in town X resulted from the text or the family background. If we had information that parental educational background enhanced knowledge retention by a factor well under 30%, we would then have reason to believe the salesman's statistics. With the information given, his statistics are at least suspect, and the conclusion is inaccurate.

(*Go to* 23)

---

23 Measurement in science is based on the assumption that numbers can be assigned to the objects, events, persons, processes, etc. being studied. These numbers correspond to the phenomenon being studied and therefore function as a symbolic representation in place of the object, event, person, process, etc. Operations possible with numbers can then be applied to this phenomenon.

However, to generalize that all such operations that are possible in our number system are possible in the phenomenon under investigation is an erroneous assumption, although it is often made. It all depends on what is being measured. For example, it is possible to assign numbers to patients in hospitals and to say that the unique number 123-785 corresponds to the unique individual, Mrs. Jones, in room 582. But it makes little sense to apply any mathematical operations to the number 123-785, which corresponds to Mrs. Jones, and the number 456-329, which corresponds to Mrs. Brown. Adding, subtracting, multiplying, or dividing will tell us nothing more about either patient. The concept being introduced is that of *scaling* or *levels of measurement*. What follows is a discussion of this concept.

(*Go to* 24)

---

24 Basic to any science is the process of classification. When elements are classified and given names, as are plants and animals, or assigned numbers, as are players on a football team, the person doing the classification must be concerned with homogeneous categories so that elements in one category are similar to each other in some way and at the same time different from elements in other categories. No comparative relationships are implied among categories. Classification is considered the lowest or most basic level of measurment. For statistical manipulation, the categories must be exhaustive (contain all the cases) and mutually exclusive (no case entered in more than one category). When we have met all of these requirements, we are using a *nominal scale.*

Identify the scale errors in these two nominal scales:

1. Physicians, interns, residents
2. Baptists, Lutherans, Methodists, Buddhists, Catholics

A. 1. Not exhaustive
   2. Not exclusive

B. 1. Not exhaustive
   2. Not exhaustive

C. 1. Not exclusive
   2. Not exclusive

D. 1. Not exclusive
   2. Not exhaustive

(A, *go to* 25) (C, *go to* 28)
(B, *go to* 27) (D, *go to* 26)

---

25 List 1: Physicians, interns, residents—is not *exclusive;* any case might be entered in more than one category because residents and interns are physicians.

List 2: Baptists, Lutherans, Methodists, Buddhists, Catholics—is not an *exhaustive* list; only a few of the religious bodies of the world are included.

Your answer was incorrect.

(*Go to* 29)

---

26 List 1: Physicians, interns, residents—is not *exclusive;* any case might be entered in more than one category because residents and interns are physicians.

In List 2, only a few of the religious bodies of the world are named, so the list is not *exhaustive*. Your answer was correct.

(*Go to* 29)

---

27 List 1: Physicians, interns, residents—is not *exclusive;* any case might be entered in more than one category because residents and interns are physicians.

List 2: Baptists, Lutherans, Methodists, Buddhists, Catholics—is not an *exhaustive* list; only a few of the religious bodies of the world are included. Your answer to this part of the question was correct.

Your answer was incorrect.

(*Go to* 29)

---

28 List 1: Physicians, interns, residents—is not *exclusive;* any case might be entered in more than one category because residents and interns are physicians. Your answer to this part of the question was correct.

List 2: Baptists, Lutherans, Methodists, Buddhists, Catholics, is not an *exhaustive* list; only a few of the religious bodies of the world are included. Your answer to this part of the question was not correct.

Your answer was incorrect.

(*Go to* 29)

---

29 It is frequently possible to *rank order* categories (that is, to put cases in numerical order from highest to lowest value) with regard to the relative amount of some characteristic that each possesses (e.g., size, value, complexity), but *not* be able to *measure* the precise amount of difference between categories or observations. This type of continuum is an *ordinal scale*. The statement $A < B < C < D$ places A, B, C, and D on an ordinal scale. ($<$ means "less than"; $>$ means "greater than.") When observations are placed on an ordinal scale, they are arranged in rank order starting with the observation having either the most or the least of the shared characteristic. Examples of ordinal scales are socioeconomic status, class ranking in school, and all kinds of judgments by observers such as severity of psychiatric impairment, degree of bone healing, demand for nursing care, or degree of family cohesiveness.

Consider the statement: The following constitute an ordinal scale: cell, tissue, organ, body.

A. Statement is false

B. Statement is true

(A, *go to* 30)
(B, *go to* 31)

---

30 If we are to arrange categories according to the degree that they possess a given characteristic, then we must place them in rank order, and we start with the observation having the most or the least of the shared characteristic. In the example we started with the smallest unit and ordered the units according to size and complexity: cell, tissue, organ, body.

(*Go to* 31)

31 The listing constitutes an ordinal scale, since we were able to place measurements in order starting with the smallest or the largest. We may actually be able to quantify the degree to which a given characteristic is possessed by each category. If we can, we now have an *interval scale*. Besides order, interval scales require that distances between categories are equal; our number system is an example. Sums and differences can be obtained from:

A. Ordinal scales
B. Interval scales

(A, *go to* 32)
(B, *go to* 33)

32 Only on the *interval* scale can we add or subtract units; your answer is not correct. Remember that on an *ordinal* scale we do not know anything about the individual values for size or weight. On an ordinal scale we may know only that A is larger than B ($A > B$). On an interval scale we may know that A is 42 and B is 10. Therefore the sum of A and B is 52, and the difference is 32.

(*Go to* 34)

33 On an interval scale we can add and subtract units; your answer is correct. If A is 42 and B is 10, an obvious sum or difference can be found.

(*Go to* 34)

34 Whether or not we actually use interval scales to establish sums or differences depends on the situation. It makes sense to add the educational attainment of a husband and wife to obtain an index value for a household educational attainment level. However, it does not make sense to add prestige of husband and wife. Many human traits do not have interval units that scientists can agree on.

Continued.

You might hear statements such as, "Student A has a class rank of 20; student B has a class rank of 10. Therefore B's class ranking is twice as high as A's."

This statement *(is/is not)* a proper use of an ordinal scale.

(*Go to* 35)

---

35 Ranks, which constitute an ordinal scale, cannot be added, subtracted, multiplied, or divided and still maintain a meaningful measure. Thus the statement is not a proper use of ordinal scales. If student A had scored 10 correct on a quiz and student B scored 20 correct, then the scores (number correct) of the two students, as an interval variable, could be added, subtracted, multiplied, or divided.

(*Go to* 36)

---

36 The fourth type of scale is the *ratio scale,* one which has an absolute zero for a starting point in addition to having equal units of measurement. The Kelvin temperature scale is a ratio scale. We can say that 400° K is twice as hot as 200° K. However, we cannot say that 100° F is twice as hot as 50° F because 0° F is not an absolute starting point. Often ratio scale examples include distances or gain scores from some starting point.

What are the four kinds of scales we may use in statistics?

(*Go to* 37)

---

37 The four types of scales are nominal, ordinal, interval, and ratio.

Exclusive, exhaustive, and lack of rank order are characteristics of the ________________ scale.

A. Nominal
B. Ordinal
C. Interval
D. Ratio

(A, *go to* 40) (C, *go to* 39)
(B, *go to* 38) (D, *go to* 41)

---

38 Although all scales have exclusive and exhaustive categories, only the nominal scale lacks rank ordering of observations. Your answer is not correct.

(*Go to* 42)

---

39 Although all scales have exclusive and exhaustive categories, only the nominal scale lacks rank ordering of observations. The interval scale, having equal units of measurement, must also include the capability of ordering. Your answer is not correct.

(*Go to* 42)

---

40 A nominal scale is the only one of the four types of scales that lacks order.

(*Go to* 42)

---

41 Although all scales have exclusive and exhaustive categories, only the nominal scale lacks rank ordering of observations. Since a ratio scale represents a higher level of mathematical sophistication than the ordinal scale, you can also rank order elements with it. Your answer is not correct.

(*Go to* 42)

---

42 Order, but not equal units, is a characteristic of the ____________ scale.

A. Interval
B. Nominal
C. Ordinal
D. Ratio

(A, *go to* 43) (C, *go to* 45)
(B, *go to* 44) (D, *go to* 46)

---

43 Only the ordinal scale has order but not equal units of measurement. Interval and ratio scales have equal units, and nominal scales do not have order.

(*Go to* 47)

---

44 The ordinal scale possesses the characteristic of order but does not have equal units of measurement. Interval and ratio scales have equal units, and nominal scales do not have order.

(*Go to* 47)

---

45 Only ordinal scales have order but not equal units of measurement. Your answer is correct.

(*Go to* 47)

---

46 Ratio and interval scales have equal units, and nominal scales do not have order. Only the ordinal scale has order without equal units of measurement.

(*Go to* 47)

---

47 Equal units and absolute zero, or absolute starting point, are characteristics of the ________________ scale.

A. Interval
B. Ordinal
C. Nominal
D. Ratio

(A, *go to* 48) (C, *go to* 50)
(B, *go to* 49) (D, *go to* 51)

---

48 Interval scales have equal units, but they do not have an absolute zero point. Only the ratio scale has *both* equal units and an absolute zero point.

(*Go to* 52)

---

49 The ordinal scale has neither equal units nor an absolute zero point. Only the ratio scale has *both* of these characteristics.

(*Go to* 52)

---

50 The nominal scale has neither equal units nor an absolute zero point. Only the ratio scale has *both* of these characteristics.

(*Go to* 52)

---

51 Equal units and an absolute zero point characterize the ratio scale. Your answer is correct.

(*Go to* 52)

---

52 Before leaving this introductory chapter, you should be clear on the concept of *variable* or *variate*. A variable is something that varies. It is an attribute for which each member of the population can be assigned a value. If every member has the same value (e.g., when using the variable "sex," every member is a male), then the attribute is called a *constant*. Since there is no information (no uncertainty) in a constant variable, we usually lose interest in such a variable for further analysis. Many times, what we actually measure is simply a *proxy*, or indicator, of the variable, or concept, in which we are interested. For example, we *operationally define* intelligence as the score on an IQ test and use the IQ score as the proxy measure for our abstract variable, or concept, called intelligence.

(*Go to* 53)

---

53 One must be clear regarding the variable in an analysis to apply an appropriate statistical technique. For example, in the following case, what is the variable, and what scale type applies?

Each day the black patients using a neighborhood health center are counted, and the count is sent to the director.

(*Go to* 54)

---

54 Race is not the variable under consideration. Race can be a variable (a nominal scale variable), but we are interested in a count of a particular racial group using the health center. That which varies is the number of black persons. The population members are the individual days of operation of the health center. A count of something is a ratio scale because the data are based on a scale of equal divisions and there is automatically a zero starting point. You must remember that the scale serves as a background against which the actual data are superimposed. The scale is not the data.

(*End of chapter* 1)

---

chapter 2

The basic comparison of numbers: ratios

At the conclusion of this chapter, you should be able to:

1. State ratios in the form A to B, A:B, and $\frac{A}{B}$ when given appropriate data.
2. State the ratio of one part to the whole, reduce the ratio to a decimal fraction (proportion), and convert the proportion to a percentage, when given a frequency table.
3. Calculate and interpret sex ratios.
4. Define and calculate a rate of change using various forms of data.
5. Understand the meaning of positive and negative rates of change.
6. State relative sizes of various ratios from comparable frequency data with different sample sizes.

1 In chapter 1 you were introduced to the four levels of measurement that are used to measure and/or classify variables. With this knowledge it is now possible to introduce some of the statistical procedures frequently found in the literature.

Since the nominal level of measurement presumes only classification of individuals with a given characteristic, it is the lowest level of measurement. However, it is always possible to engage in the following mathematical operations or calculations with a nominal scale or any higher level of measurement: ratios, proportions, and percentages.

Ratios are often used to demonstrate the numerical relationships of two measurements. Generally, these two measurements come from the same population or sample. A ratio is sometimes expressed in the form of a fraction to demonstrate the relationship between the two parts of the population or sample. For example, the ratio of A to B is written as the fraction $\frac{A}{B}$.

(*Go to* 2)

---

2 Of 76 women having undergone induced ovulation, 29 of the pregnancies resulted in multiple births. Of these 29 multiple-birth pregnancies, 23 produced twins. Of the twins, there were 2 male-male sets, 10 male-female sets, and 11 female-female sets (Hack et al., 1970). The ratio of mixed (male-female) sets of twins to total sets of twins born is:

A. $\frac{23}{10}$

B. $\frac{10}{23}$

(A, *go to* 4)
(B, *go to* 3)

---

3 The first-named term of a ratio is always the numerator (top number) of the fraction; therefore your answer, $\frac{10}{23}$, is the correct ratio of mixed to total sets of twins.

(*Go to* 5)

---

4 The ratio $\frac{23}{10}$ is the ratio of total sets of twins to mixed-sex sets of twins.

Your answer is incorrect. The ratio of mixed to total sets of twins is $\frac{10}{23}$. The first-named term of a ratio is always the numerator (top number) of the fraction.

(*Go to* 5)

---

5 If the ratio of mixed sets to total sets of twins born after induced ovulation is $\frac{10}{23}$, what is the sex ratio (males to females) of all twin children born in the sample after this medical procedure, where 2 sets are male-male, 10 sets are male-female, and 11 sets are female-female?

Write your answer before proceeding to the next frame.

(*Go to* 6)

---

6 You should have determined the total number of male and female infants born; that is, 2 sets of male-male twins = 4 males, 10 sets of male-female twins = 10 males and 10 females, and 11 sets of female-female twins = 22 females—or for males, 4 + 10 = 14, and for females, 10 + 22 = 32 females. The sex ratio is therefore 14 males to 32 females, or $\frac{14}{32}$.

Usually a constant of 100 is incorporated into the sex ratio formula, or $\frac{\text{number of males}}{\text{number of females}} \times 100$. Therefore for the above sample of induced-ovulation births, the sex ratio for *twins* is $\frac{14}{32} \times 100 = 44$ male births for every 100 female births. The *overall* sex ratio at birth in the United States is about 104; that is, 104 males for every 100 females born. The method of induced ovulation reported by Hack et al. appears to have had some effect on the sex ratio of twins.

(*Go to* 7)

---

7 Given the data in Table 2-1, what is the ratio of middle-aged Catholic faculty to younger Catholic faculty?

(*Go to* 8)

TABLE 2-1. Numbers of faculty, by age and religious background, of "elite" colleges and universities, 1969*

Age (years)	*Religious background*		
	Protestant	*Catholic*	*Jewish*
55 and over (older)	1996	264	326
40-54 (middle-aged)	4063	811	1392
39 and under (younger)	5024	1516	1906
TOTAL	11,083	2591	3624

*Modified from Greeley, A. M. 1973. The "religious factor" and academic careers: another communication. Am. J. Soc. **78:**1247-1255.

8 The ratio of middle-aged to younger Catholic faculty at "elite" colleges and universities is $\frac{811}{1516}$.

(*Go to* 9)

9 What is the ratio of younger Protestant faculty to younger Jewish faculty from Table 2-1?

(*Go to* 10)

10 The ratio of younger Protestant to younger Jewish faculty at "elite" colleges and universities is $\frac{5024}{1906}$.

(*Go to* 11)

11 Ratios may be expressed in other forms than the fraction form we have been using. A ratio can be stated in words. Another ratio expression is of the form A : B, where the first-named term always precedes the colon. Thus the ratio of 1996 older Protestant faculty to 326 older Jewish faculty written in this form would be ______________.

Write your answer before going on.

(*Go to* 12)

12 This expression of the ratio is 1996:326. Express the ratio X to Y in the two forms.

Write your answer before going on.

(*Go to* 13)

13 The ratio X to Y may be written $\frac{X}{Y}$ or X:Y.

(*Go to* 14)

14 To more easily visualize the relationship between two numbers, it is common practice to reduce ratios to the smallest fraction. The ratio $\frac{12}{27}$ is much harder to comprehend than the reduced form of $\frac{4}{9}$. The reduced form was arrived at by dividing each part of the ratio by 3 ($12 \div 3 = 4$; $27 \div 3 = 9$).

Hong (1972) conducted a study of kinship in rural and urban Hong Kong. In one of his analyses he used 272 rural respondents and 2672 urban respondents. The ratio of rural to urban respondents is:

A. $\frac{2672}{272}$

B. $\frac{167}{17}$

C. $\frac{17}{167}$

D. None of these

(A, *go to* 15) (C, *go to* 17)
(B, *go to* 16) (D, *go to* 18)

15 Your answer, $\frac{2672}{272}$, is not correct. You have the ratio of urban to rural respondents. Furthermore, you did not reduce it to its lowest terms. The correct answer is $\frac{272 \div 16}{2672 \div 16} = \frac{17}{167}$, or 17 rural to 167 urban respondents.

(*Go to* 19)

16 You have reduced the individual terms to the lowest numbers possible, but you have presented the ratio of urban to rural respondents. The question asked for rural to urban respondents; the correct answer is $\frac{17}{167}$.

(*Go to* 19)

17 Your answer is correct.

(*Go to* 19)

---

18 There are two possibilities for the error that you just made:

1. Your math may be incorrect.
2. You may not have reduced the ratio to the lowest possible terms.

Try that frame again.

(Go to 14)

---

19 It is common to express a ratio in terms of a denominator of 1; e.g., 2:1, 8 to 1. Reducing the ratio of $\frac{12}{4}$ or 12:4 to 3:1 is easily accomplished by dividing both terms by the denominator; i.e., 4.

However, if the first term is not evenly divisible by the second, a decimal results; e.g., reducing the ratio of 127:7 results in 18.1 to 1, or 18.1:1.

Express the ratio 288:7 as a ratio of ______________ to 1.

Write your answer before continuing.

(Go to 20)

---

20 288:7 reduces to a ratio of 41.1 to 1 (because $\frac{288}{7} = 41.1$).

Kelman and Lawrence (1972), in a sample of 989 persons over 18 years of age, asked questions regarding the trial of a soldier accused of a massacre during the Vietnam War. They found that 574 disapproved of the trial, 336 approved, and 79 had no opinion. Express the ratio of those disapproving to those approving:

1. As a fraction ______________
2. Not as a fraction ______________
3. In the form ______________ to 1

Perform your calculations before going on.

(Go to 21)

---

21 The ratio 574 to 336 (disapproving to approving) can be written:

1. As a fraction: $\frac{574}{336}$, or reduced to $\frac{287}{168}$
2. Not as a fraction: 574:336, or reduced to 287:168
3. In the form 1.7 to 1 or 1.7:1

The 1.7:1 form of the ratio expresses most clearly the relationship between the two categories in this case; 1.7 persons disapprove for every 1 person who approves of the trial.

Continued.

Since "1.7 persons" is a cumbersome manner of speaking of whole people, it would be appropriate to multiply both the 1.7 and the 1 by some larger number. For example, it is perhaps easier to understand the statement that 17 persons disapprove for every 10 who approve of the trial ($1.7 \times 10 = 17$; $1 \times 10 = 10$).

Other population bases such as 100, 1000, or 100,000 may be used to keep the numerator of the ratio from being less than 1. The critical factor here is the ease of understanding the resulting statistic.

(*Go to* 22)

---

22 The special type of ratio in which the denominator is the total number of cases and the numerator is some part of the total is called a *proportion.*

Hurst (1972), using data from random samples of "middle-sized United States cities," found that among 281 low-occupational-status blacks, 172 perceived a high degree of economic subjugation for black people as a whole and 98 perceived a medium degree. Of 122 high-occupational-status blacks, 58 perceived a high degree and 56 a medium degree of economic subjugation. (The remainder in each occupational category perceived no economic subjugation.) The *proportion* of low-occupational-status blacks perceiving a high degree of economic subjugation is 0.61 ($\frac{172}{281} = 0.61$).

Of all blacks interviewed, what proportion perceived a medium amount of economic subjugation?

A. 0.35
B. 0.38
C. None of the above

(A, *go to* 23)
(B, *go to* 24)
(C, *go to* 25)

---

23 Your answer represents the proportion of low-occupational-status blacks perceiving a medium amount of economic subjugation: $\frac{98}{281} = 0.35$. The question was, for all blacks in the sample, what proportion perceived a medium amount of economic subjugation?

$$\frac{98 + 56}{281 + 122} = \frac{154}{403} = 0.38$$

(*Go to* 26)

---

24 Your answer is correct.

(*Go to* 26)

---

25 For all blacks in Hurst's sample, the proportion perceiving a medium amount of economic subjugation is $\frac{98}{281}$ (for low occupational status) + $\frac{56}{122}$ (for high occupational status) $= \frac{154}{403} = 0.38$.

(*Go to* 26)

---

26 Proportions are often expressed as percentages (%) rather than as decimals simply by multiplying the decimal or proportion by 100. Thus two entirely equal statements can be made.

1. The *proportion* of all blacks perceiving a medium amount of economic subjugation is 0.38.

2. The *percentage* of all blacks perceiving a medium amount of economic subjugation is 38%.

Convert the following proportion to a percentage:

A proportion of 0.325 = ____________%.

Write your answer before going to the next frame.

(*Go to* 27)

---

27 Your answer should be 32.5%. Always move the decimal two places to the right in converting proportions to percentages.

Because *percent* means *per hundred*, a percentage measure automatically indicates the incidence of observations (persons, cases, outcomes, etc.) per hundred population. Thus 32.5% indicates 32.5 observations per hundred in the population.

Another measure frequently used is a type of ratio called *rate of increase* or *rate of change*. This rate results from the following calculation:

$$\text{Rate of increase (change)} = \frac{\text{Measurement now} - \text{Measurement then}}{\text{Measurement then}}$$

Millimet and Gardner (1972) conducted an experiment using an inventory of feelings (affect) score as influenced by a successful experimental outcome compared to an experimental outcome that was a failure. They found that for respondents with low self-esteem, the affect score changed from 134.13 before the successful experimental outcome to 150.27 after the same experimental setting. What is the rate of change in this score?

A. 0.89 or 89% (A, *go to* 28)

B. 0.12 or 12% (B, *go to* 29)

C. 0.107 or 10.7% (C, *go to* 30)

---

28 Inspection should indicate that a change from an affect score of 134.13 to 150.27 could not be 89%, or nearly double. You must have arrived at 89% by dividing 134.13 by 150.27. The calculation (or formula) for rate of change is:

$$\frac{\text{Measurement now} - \text{Measurement then}}{\text{Measurement then}}$$

Solving:

$$\frac{150.27\,(\text{now}) - 134.13\,(\text{then})}{134.13\ (\text{then})} = \frac{16.14}{134.13} = 0.12$$

or 12%. There was a 12% change in the affect score for persons with low self-esteem after a successful experiment.

(*Go to* 31)

---

29 Your answer, 0.12 or 12%, is the correct rate of change between the two times. In other words, there was a 12% change in the affect score for low self-esteem persons after a successful experiment.

(*Go to* 31)

---

30 Your answer, 0.107 or 10.7%, is incorrect. Your answer was found by calculating:

$$\frac{150.27 - 134.13}{150.27}$$

Rate of change =

$$\frac{\text{Measurement now} - \text{Measurement then}}{\text{Measurement then}}$$

Solving:

$$\frac{150.27\,(\text{now}) - 134.13\,(\text{then})}{134.13\ (\text{then})} = \frac{16.14}{134.13} = 0.12 \text{ or } 12\%$$

There was a 12% change in the affect score for low self-esteem persons after a successful experiment.

(*Go to* 31)

31 A rate of change may be negative, showing a decrease rather than an increase.

In the same experiment reported earlier, Millimet and Gardner used high self-esteem respondents and placed them in an experimental setting that failed. Their affect score went from 160.2 before the experiment to 154.6 after the failure. What is the rate of change?

A. 0.035 or 3.5% (A, *go to* 32)
B. −0.035 or −3.5% (B, *go to* 33)
C. −0.036 or −3.6% (C, *go to* 34)

---

32 Your answer, 0.035 or 3.5%, is incorrect. There was a *decrease* in affect for the high self-esteem group after the failure of −3.5%. The negative value must be in the answer to indicate a decrease.

(*Go to* 35)

---

33 Your answer, −0.035 or −3.5%, is correct. The negative value indicates that affect for the high self-esteem group decreased by −3.5% after failure.

(*Go to* 35)

---

34 You calculated $\frac{154.6 - 160.2}{154.6}$. The denominator should be the original value, not the most recent value. The correct answer is −0.035 or −3.5% and was arrived at as follows:

$$\frac{154.6\ \text{(now)} - 160.2\ \text{(then)}}{160.2\ \text{(then)}} = -\frac{5.6}{160.2} = -0.035 \text{ or } -3.5\%$$

The negative value indicates that affect for the high self-esteem group decreased −3.5% after failure.

(*Go to* 35)

---

35 Table 2-2 comes from a review article by Stein and Susser (1970), who cite research in England on the university grades of students according to father's occupation. It is difficult to see the relationship between grades and father's occupation because of the difference in the size of the two groups of occupations (803 vs. 94). For example, the number of "above average passes" either lead or are at least tied for first place with other grades in the two occupational groups or populations. However, are these grades of the same relative statistical importance in the two groups?

Continued.

TABLE 2-2. Grades of university students by father's occupational level (number in each category)*

Grade	Father's occupation: Nonmanual	Father's occupation: Semiskilled and unskilled manual
Superior pass	209	23
Above average pass	233	36
Ordinary pass	233	26
No degree (failed)	128	9
TOTAL	803	94

* Modified from Stein, Z., and M. Susser. 1970. Mutability of intelligence and epidemiology of mild mental retardation. Rev. Educ. Res. **40**:29-67 (Feb.).

Table 2-2 can be made more meaningful by using one of the measures we have been discussing. For example, the proportion of "ordinary passes" among nonmanual father's occupation students is $\frac{233}{803} = 0.29$, whereas for the children of semiskilled and unskilled manual workers, the proportion of "ordinary passes" is $\frac{26}{94} = 0.277$. Although the nonmanuals have over 200 more "ordinary passes," the difference in proportion between them and the semiskilled and unskilled manuals is only 0.013.

The phrase "only 0.13" has mathematical implications only, not substantive meaning. The person doing the research and the reader must decide whether the mathematical answer provided by statistics has any substantive meaning.

Compare the "above average passes" in the two occupational groups using proportions. Make your calculations before you go on.

(*Go to* 36)

---

36 The proportion of "above average passes" among the nonmanuals is $\frac{233}{803} = 0.29$, whereas for the semiskilled and unskilled manual workers' children, it is $\frac{36}{94} = 0.383$. Although the children of nonmanually occupied fathers have almost 200 more grades of "above average pass" than do the other group, the nonmanuals have a lower proportion of the total grades in the "above average pass" category.

(*Go to* 37)

---

37 Because a percentage is a proportion multiplied by 100, it also can be used to interpret Table 2-2. Use percentages to compare the number of "failing" or "no degree" grades among the children of the two occupational groups. Make all of your calculations before going on.

(*Go to* 38)

TABLE 2-3. Grades of university students by father's occupational level*

	Father's occupation			
	Nonmanual		*Semiskilled and unskilled manual*	
Grade	N	%	N	%
Superior pass	209	26.0	23	24.5
Above average pass	233	29.0	36	38.3
Ordinary pass	233	29.0	26	27.7
No degree (failed)	128	15.9	9	9.6
TOTAL	803	99.9	94	100.1

*Modified from Stein, Z., and M. Susser. 1970. Mutability of intelligence and epidemiology of mild mental retardation. Rev. Educ. Res. **40**:29-67 (Feb.).

38 Among the nonmanuals, $\frac{128}{803} \times 100 = 15.9\%$ of the university students failed, whereas for the children of semiskilled and unskilled manual workers, only $\frac{9}{94} \times 100 = 9.6\%$ failed.

Table 2-3 is a modification of Table 2-2 showing the percentages for each grade achieved. Note how much more useful Table 2-3 is than was Table 2-2. Note also that rounding answers to the nearest tenth of a percent makes a total of 99.9% in one case and 100.1% in the other. More accurate calculations, that is, calculations carried to more decimal places, should result in an even 100% in each case. In addition, it is important to include the actual number of persons (observations) contributing to each percentage calculation. This provides additional information for the reader.

(*Go to* 39)

---

39 If we wanted to achieve 100% in Table 2-3, we could do this by adding or subtracting 0.1% to any one of the percentages. From which grade among the semiskilled and unskilled manual workers would you subtract 0.1%?

A. "No degree" (9.6%)

B. "Above average pass" (38.3%)

(A, *go to* 40)
(B, *go to* 41)

---

40 By subtracting the additional 0.1% from the 9.6% of the "no degree" people we would be affecting the smallest category. The least error is introduced by subtracting the 0.1% from the *largest* category. In other words, change the "above average pass" category from 38.3% to 38.2% to avoid the problem of stating a "total" of greater than 100%.

(*Go to* 42)

---

41 We would introduce the least error by subtracting 0.1% from the grade category having the largest proportion, i.e., "above average pass." The 0.1% is a smaller part of 38.3% than of any of the other values, and, therefore, it introduces less error when subtracted from the largest percentage.

In other words, change the "above average pass" category from 38.3% to 38.2% to avoid the problem of stating a "total" of greater than 100%.

(*Go to* 42)

---

42 To which category in the nonmanual father's occupation group would you add 0.1% to achieve a total of 100%: the "above average pass" or the "ordinary pass" category?

Make your decision before going on.

(*Go to* 43)

---

43 The answer to that question is anybody's guess! Your best bet in this situation is to flip a coin or leave the total at 99.9%.

(*End of chapter* 2)

---

chapter

3

Frequency distributions and graphics

At the conclusion of this chapter, you should be able to:

1. State the range of a given array of data.
2. Given a data set, construct a frequency table involving intervals and exhibiting the recommended properties of a frequency table (interval range small enough to show detail, not more than 15 intervals, odd-whole number interval ranges, starting value a multiple of the interval size).
3. Compute the values for cumulative frequency, cumulative proportion, and cumulative percentage.
4. Given a frequency distribution table, write correct interpretive statements, involving frequencies, proportions, and percentages.
5. Given an appropriate data set, construct and interpret a bar graph.
6. Given a frequency table with data in intervals, construct, label, and interpret a frequency polygon and a histogram.
7. Develop an understanding of the concept "skewness" (referring to X axis nonsymmetry) and demonstrate this understanding by (a) being able to label correctly a curve as positively or negatively (or right or left) skewed and (b) being able to sketch a curve to illustrate positive or negative skewness.
8. Develop an understanding of the concept "kurtosis" (referring to Y axis peakedness) and demonstrate this understanding by (a) being able to label correctly a curve as platykurtic or flat, leptokurtic or pointed, or mesokurtic or normal, and (b) being able to sketch a curve to illustrate each kurtosis previously named.
9. Given a frequency distribution table with data in intervals, construct and correctly label a cumulative proportion (or cumulative percentage) curve (ogive curve).

1 As mentioned in previous chapters, one of the purposes of statistics is to reduce large masses of data into more manageable and interpretable forms. The use of frequency distributions and graphic presentations is one step in this process.

For any given set of data, the first step in organizing a frequency distribution is to determine the *range* of scores or measurement values. This is easily done by subtracting the lowest score (or measure) from the highest score in the set of scores and then adding one; that is, $(X_h - X_l) + 1 =$ Range, where X_h is the raw score of the highest value and X_l is the raw score of lowest value.

The range for the set of scores on a depression scale presented in Table 3-1 would be:

A. 55 (A, *go to* 2)
B. 54 (B, *go to* 3)

TABLE 3-1. Raw scores obtained from a sample of mental patients on a standardized scale measuring depression

56	78	62	37	54	39	62	60
28	82	38	72	62	44	54	42
42	55	57	65	58	47	42	56
56	56	55	66	42	52	48	48
47	41	50	52	47	48	53	68

2 The lowest score in the set is 28 and the highest is 82. Therefore the range is found as follows: $82 - 28 + 1 = 55$. Your answer is correct.

(*Go to* 4)

---

3 The lowest score in the set is 28 and the highest score is 82. $82 - 28 = 54$. However, the number 1 must be added to the difference because the formula is $(X_h - X_l) + 1 =$ Range. Therefore $(82 - 28) + 1 = 55$.

(*Go to* 4)

---

4 A good deal more will be said about the statistic called the range in a later chapter when the concept of dispersion is introduced. The range is presented here since it will be helpful in determining the size of the *interval* unit used in the *frequency table*. Intervals in frequency tables are ways of classifying raw data into smaller categories. The size of the interval is always related to the data. For example, in blood pressure, which is measured in individual units called millimeters of mercury (mm Hg), an interval size of 50 mm would be so large as to be meaningless. Blood pressure would more appropriately be presented in units or intervals of 5 mm. On the other hand, the distribution of incomes in a given community would probably best be presented in units of $1000 or more.

Using the data on depression scores presented in Table 3-1, we are interested in establishing a *frequency distribution* and must decide on an appropriate number of intervals. In this example, 15 might be a reasonable number of intervals, since it will give us an interval unit or *size of interval* that makes sense in interpreting depression scores.

By dividing the range by the number of intervals we can get the size of the interval. In the case of the cited depression scores, with a range of 55, the interval size would be:

A. 3
B. 4
C. Between 3 and 4

(A, *go to* 5)
(B, *go to* 6)
(C, *go to* 7)

---

5 The division of $\frac{55}{15}$ would give an answer of between 3 and 4. However, the interval size is generally a whole number, so either 3 *or* 4 could be used. It is more convenient, as will become apparent, to use an odd number for the interval size because the midpoints of each of the intervals will be whole numbers. Thus numbers such as 3 and 5 are commonly used for interval size. In this case, 3 would be our choice.

(*Go to* 8)

---

6 Your answer was 4. The correct answer is between 3 and 4. However, the interval size is usually a whole number; therefore 4 could be used. It is more convenient, as will become apparent, to use an odd number for the interval size because the midpoints of each of the intervals will then be whole numbers. Thus numbers such as 3 and 5 are commonly used for interval size. In this case, 3 would have been a better choice.

(*Go to* 8)

---

7 Since you have chosen "between 3 and 4" as your answer, your division is correct. However, the interval size is usually a whole number so either 3 or 4 could be used. It is more convenient, as will become apparent, to use an odd number for the interval size because the midpoints of each of the intervals will be whole numbers. Thus numbers such as 3 and 5 are commonly used for interval size. In this case, 3 would be our choice.

(*Go to* 8)

---

8 Now that the interval size has been determined, we can start constructing the interval calibration. The determination of where to start is done by allowing the lowest interval to begin with a number that is a multiple of the interval size; that is, the lowest number in the lowest interval should be able to be arrived at by multiplying the interval size by some other number. This number must be either the same or lower than the lowest value in the array of raw numbers. Since with the example we have been using, the lowest score was 28, the lowest interval should be:

A. 28-31
B. 27-29
C. 29-32

(A, *go to* 9)
(B, *go to* 10)
(C, *go to* 11)

---

9 Although the interval that you chose, 28-31, will include the lowest measure in the set, it does not start with a multiple of 3. A better interval would be 27-29, since 27 is divisible by 3.

(*Go to* 12)

---

10 Your answer, 27-29, would be the most appropriate lowest interval because 27 is divisible by 3, and the interval includes the lowest measure in the set.

(*Go to* 12)

---

11 The interval 29-32 would not include the lowest measure in the set, nor is 29 divisible by 3. The best interval would be 27-29, since it would include 28, the lowest measure, and since 27 is a multiple of 3.

(*Go to* 12)

---

12 After finding the lowest interval, interval calibration is simply a matter of listing the intervals in a column until the highest interval is reached. The highest interval is the one that is a multiple of the interval size and includes the highest measure in the set of values.

Construct a frequency table to represent the data on depression scores found in Table 3-1. The first interval should be 27-29, and the last interval should contain the largest score in the array of scores (82). Each interval should be a multiple of 3. Then tally each of the scores to be found in the array and place them in the appropriate intervals. Your table should be in the following form:

Interval	*Tally*	*Frequency (f)*
.		
.		
.		
.		
.		
27-29	I	1

When you are done, enter the total in the "frequency" column by adding up all frequencies. Complete this on a separate sheet of paper, and then compare your work with the completed frequency table (Table 3-2). Save your work, since you will use it again.

(*Go to* 13)

TABLE 3-2. Frequency distribution of standardized depression scale for sample of mental patients—I

Interval	*Tally*	*Frequency (f)*
81-83	I	1
78-80	I	1
75-77		0
72-74	I	1
69-71		0
66-68	II	2
63-65	I	1
60-62	IIII	4
57-59	II	2
54-56	~~IIII~~ III	8
51-53	III	3
48-50	IIII	4
45-47	III	3
42-44	~~IIII~~	5
39-41	II	2
36-38	II	2
33-35		0
30-32		0
27-29	I	1
TOTAL		40

13 Note that in the frequency distribution that you constructed there are 19 separate intervals. Although we originally selected 15 intervals, the fact is that $\frac{55}{15}$ (the range divided by 15 intervals) resulted in an interval size of 3 to 4. Thus there are slightly more than the 15 intervals in your frequency distribution.

Examine the data organized in the frequency distribution that you prepared. Note that the preponderance of the 40 scores fell in the low-to-middle end of the distribution; that is, the mental patients, for the most part, had depression scores between about 42 and 62 in a possible range of 55 points. In this way the frequency distribution has served to represent an array of 40 numbers in a more meaningful and more easily analyzable manner.

(*Go to* 14)

14 Summary of steps in constructing a frequency distribution table:

1. Determine the range.
2. It is customary to divide the range by a number equal to or less than 15. Any number in excess of 15 makes the resulting table difficult to read or interpret. The most important point is that the size of the *interval* selected be *revelant to the data.*
3. List the intervals in a column beginning at the bottom. Let the lowest interval begin with a number that is a multiple of the interval size.
4. Tally the frequencies.
5. Summarize these under the "frequency" (f) column.
6. Total the "frequency" column and record the number of cases at the bottom.

(*Go to* 15)

15 Frequency distributions, with their associated percentages, are used often in the research literature of any field. They represent a straightforward and simple manner of displaying the relationships between various categories of a variable. Zimet et al. (1971) analyzed instructional books (primers) for young children for differences in content in 13 different countries. They presented percentage distributions in which the American primers were shown to have the most "other-directed content" (dealing with interpersonal interaction) of all of the countries, whereas Israel displays the highest "culture" orientation (i.e., dealing with "collective symbols" of the country's people).

(*Go to* 16)

16 After determining the intervals and frequency for each interval, another column can be placed next to the frequency column. This will indicate the *cumulative frequency,* or the number of cases at that interval plus the number of cases below (or above) the given interval. For finding the value in each interval of this column, note the number of scores or measures *preceding* the upper limit of that interval. Therefore, in the example just completed, if you start at the bottom interval (27-29), the number of occurrences below the upper limit would be 1. For the fourth from the bottom interval (36-38), the cumulative frequency would be 3, since 3 scores occurred below 38.

Calculate the cumulative frequency for the data presented in Tables 3-1 and 3-2 starting from the lowest score. The symbol generally used for the cumulative frequency column is "F." When you have finished your work, check it in Table 3-3.

(*Go to* 17)

17 Examining your work indicates that up to the score 56, which is the highest limit of the interval containing the greatest proportion of scores, 28 of the 40 patients' scores can be represented; that is, 28 scores are less than or equal to 56. Although this is useful information in an attempt to reduce the original array of data to a more easily interpretable form, the use of proportions would make the data even more meaningful. The advantage of using proportions or percentages is that the base or total to which you relate a number is always the same (1.0 for proportions, 100 for percents), regardless of the number of cases being counted.

Therefore the next step is to obtain the *cumulative proportions and percentages*. This is done by dividing the cumulative frequency of each interval by the total number of scores. Thus in our example the cumulative proportion for the first interval would be:

$$\frac{\text{Cumulative frequency}}{\text{Total N}} = \frac{1}{40} = 0.025 = \text{Cumulative proportion}$$

To change the cumulative proportion to cumulative percentages, multiply the value of the cumulative proportion by 100. Therefore $0.025 \times 100 = 2.5\%$.

What are the cumulative proportion and cumulative percentage for the interval 54-56 in the array of data on depression scores? Calculate this number and then check it in Table 3-3.

(*Go to* 18)

TABLE 3-3. Frequency distribution of standardized depression scale for sample of mental patients—II

Interval	*Tally*	*Frequency* (*f*)	*Cumulative frequency* (*F*)	*Cumulative proportion*	*Cumulative percentage*
81-83	I	1	40	1.00	100%
78-80	I	1	39	0.975	97.5%
75-77		0	38	0.95	95%
72-74	I	1	38	0.95	95%
69-71		0	37	0.925	92.5%
66-68	II	2	37	0.925	92.5%
63-65	I	1	35	0.875	87.5%
60-62	IIII	4	34	0.85	85%
57-59	II	2	30	0.75	75%
54-56	~~IIII~~ III	8	28	0.70	70%
51-53	III	3	20	0.50	50%
48-50	IIII	4	17	0.425	42.5%
45-47	III	3	13	0.325	32.5%
42-44	~~IIII~~	5	10	0.25	25%
39-41	II	2	5	0.125	12.5%
36-38	II	2	3	0.075	7.5%
33-35		0	1	0.025	2.5%
30-32		0	1	0.025	2.5%
27-29	I	1	1	0.025	2.5%

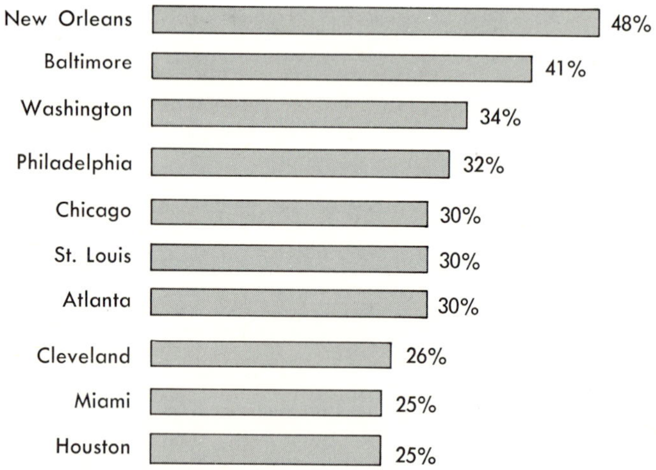

FIG. 3-1. Percent total public elementary school enrollment that is nonwhite, for selected urbanized areas, 1970. (Modified from Hermalin, A. I., and Farley, R.: Am. Sociol. Rev. **38:**595-610, 1973.)

18 Referring to Table 3-3, we now know that 70% of the sampled mental patients had depression scores of 56 or less. Each interval must have its cumulative percentage calculated so that an *ogive* (pronounced ō′jīv) *curve* can be constructed. More will be said about the ogive curve in the following discussion on graphing.

Would you guess the graphic illustration, Fig. 3-1, to be a:

A. Frequency polygon (A, *go to* 19)
B. Bar graph (B, *go to* 20)

---

19 The graph is not a *frequency polygon;* it is a *bar graph.* We will talk more about frequency polygons later. For the time being let us look more closely at why this is an example of a bar graph.

The bar graph is used to represent data that is either *nominally* or *ordinally* scaled. The bar graph illustrating the percentage of the total public elementary school enrollment that was nonwhite is based on which of the following levels of measurement?

A. Interval (A, *go to* 21)
B. Ordinal (B, *go to* 22)
C. Nominal (C, *go to* 23)

---

20 You called the graph a *bar graph,* which is correct. It has a number of characteristics that distinguish it from a *frequency polygon* or a *histogram.*

The bar graph is used to represent data that are either *nominally* or *ordinally* scaled. The bar graph illustrating the percentage of the total public elementary school enrollment that was nonwhite is based on which of the following levels of measurement?

A. Interval (A, *go to* 21)
B. Ordinal (B, *go to* 22)
C. Nominal (C, *go to* 23)

---

21 The level of measurement used here is not interval. Although the urbanized areas were listed in descending order of nonwhite enrollment, it was not necessary to order this way. The urbanized areas could have been arranged in alphabetical order, and it would have been just as proper. In a bar graph various categories can be represented in any order you choose. The correct answer is that this bar graph represents a nominally scaled variable. (*Go to* 24)

---

22 The graph does not depict an ordinal level of measurement. You might have been misled by the fact that the categories were depicted in decreasing percentages of nonwhite enrollment. However, the categories could just as well have been ordered by an alphabetical listing of urbanized areas. There is no rank ordering of the data, and the graph more properly represents a nominally scaled variable.

(*Go to* 24)

---

23 The graph depicts a nominally scaled variable. Although the urbanized areas were listed in descending order by percentage of nonwhite enrollment, it was not necessary to order this way. The categories could have been arranged in alphabetical order, and it would have been just as proper. In a bar graph the various categories can be represented in any order you choose.

(*Go to* 24)

---

24 An examination of the bar graph represented in Fig. 3-1 indicates that of the 10 listed urbanized areas, New Orleans has the greatest proportion of its public elementary school enrollment categorized as nonwhite (48%), whereas Miami and Houston, of the listed urbanized areas, have the lowest percentage of nonwhite enrollment (25%).

Another feature of Fig. 3-1 that is critical is that the bars for each urbanized area are separated rather than continuous. The separation avoids any implication of continuity among the categories. Because the graph pictorially describes the data, the bars also:

A. Should all be the same color (A, *go to* 25)

B. Should all be the same thickness (B, *go to* 26)

---

25 The purpose of the bar graph is to visually display a measure of quantity. A variation in color from one bar to the next will not, normally, affect or imply differences in amount. Color variation may, however, distinguish differences in categories, which would be a possible advantage. If the bars were of different thicknesses, the implication may be made that there are differences in amount. Characteristically, the length of the line is the indicator of amount or quantity. It is therefore best to keep the thickness of bars constant.

(*Go to* 27)

---

26 Differences in thickness of the bars may imply differences in amount or quantity. Therefore it is necessary to keep the bar thicknesses the same. The length of the bar is the indicator of size, amount, or whatever unit is to be represented, since each bar represents the proportion of *total area* devoted to that category.

(*Go to* 27)

---

27 Bar graphs, like frequency distributions, are found often in the literature. They represent an easily interpretable manner of presenting findings. Born and Herbert (1971) present a bar graph of grades earned in a college course using an experimental teaching technique that was designed to be highly personalized. They argue from their bar graph that students did better in the personalized technique than they did in normal large classroom settings.

Whitman (1969) found in a free recall situation that subjects who were taught to memorize a list of words as opposed to those who were taught to understand a rule for the organization of the words to be recalled later did less well in remembering a subsequent list of words for a free recall situation. Whitman presents his data in a bar graph that makes understanding of his findings straightforward and easily interpretable.

(*Go to* 28)

---

28 The graph presented in Fig. 3-2 is an example of a *frequency polygon.* It illustrates the frequency distribution of systolic blood pressures for a sample of 9- and 10-year-old children. Each line shows the frequency distribution for one age group. The data presented indicate that for all of the measurements made, most of the systolic blood pressures are concentrated between about 90 and 120.

(*Go to* 29)

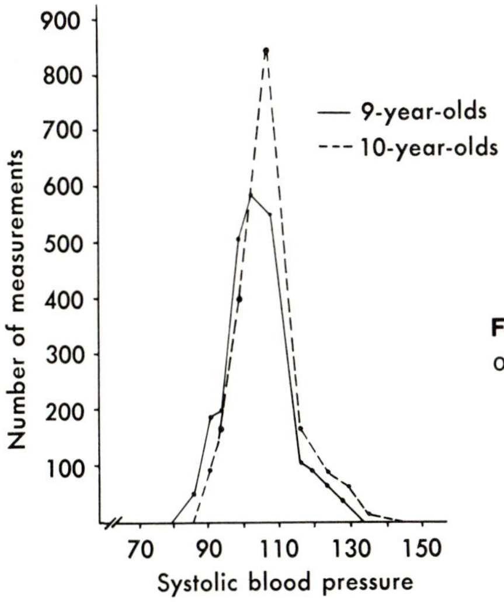

FIG. 3-2. Systolic blood pressure for samples of 9- and 10-year-old children.

29 To construct a frequency polygon from one or more frequency distributions, first lay out an X and a Y axis. The X axis is a horizontal line that intersects with the vertical Y axis as shown in Fig. 3-3. The point where they meet is the zero point for the scales running along either axis. This starting point is called the:

A. Abscissa
B. Origin

(A, *go to* 30)
(B, *go to* 31)

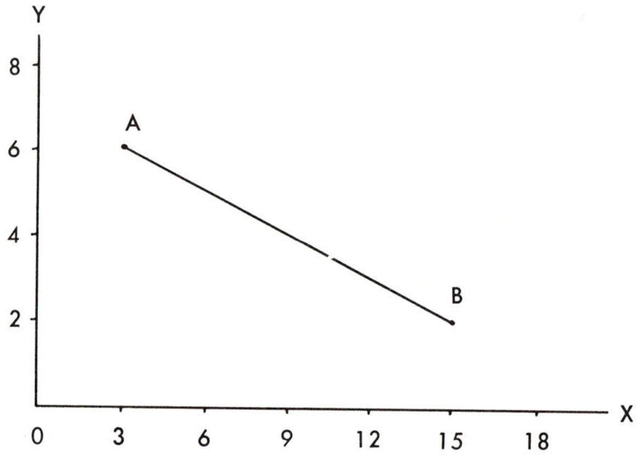

Fig. 3-3. Frequency polygon for variable X.

30 The abscissa is the horizontal line or X axis. The correct answer is the *origin,* or the point at which the X axis meets the Y axis at zero. The name given to the Y axis is the *ordinate.*

(*Go to* 32)

31 The *origin* is the starting point for the scales of both *abscissa* and *ordinate* values. The abscissa refers to the X axis, and the ordinate refers to values of the Y axis.

(*Go to* 32)

32 As you may have noticed in the example, the Y axis (ordinate) was used to indicate the *frequency* of measures, whereas the X axis (abscissa) indicated the *values* of the measurement scales, or the variable under consideration. Furthermore, the two scales must be calibrated to:

A. Include the lowest and highest frequency for any one value of the measurement scale
B. Include the highest and lowest values measured in the sample
C. Both of the above

(A, *go to* 33)
(B, *go to* 34)
(C, *go to* 35)

---

33 Not only must the Y axis be calibrated to include the highest frequency for any one score, but the X axis must also be calibrated to include the highest value measured in the sample. Your answer was incorrect.

(*Go to* 36)

---

34 Not only must the X axis be calibrated to include the highest value measured in the sample, but the calibration of the Y axis must also be high enough to include the highest frequency for any one measurement value. Your answer was incorrect.

(*Go to* 36)

---

35 Your answer is correct; you must calibrate on the X axis for the highest value that was measured for any one individual. You must likewise calibrate the Y axis for the highest:

A. Value of the *independent variable*
B. Frequency value obtained for any one value along the X axis

(A, *go to* 37)
(B, *go to* 38)

---

36 The highest value of the ordinate scale will be:

A. The highest value on the X axis scale
B. The highest frequency of occurrence of the measure

(A, *go to* 37)
(B, *go to* 38)

---

37 The *ordinate* scale is the scale of values along the Y axis (vertical axis) and in a frequency polygon is used to indicate *frequency*. Therefore the highest value on the ordinate scale will be the highest frequency obtained for any one value on the X axis, or measurement scale. Your answer was not correct.

(*Go to* 39)

---

38 The ordinate scale denotes the frequencies of each value represented on the X axis. Therefore the highest value of the Y axis will be the highest frequency of occurrence for any one value on the X axis (abscissa), or measurement scale. Your answer was correct.

(*Go to* 39)

39 The highest frequency for any one value on the abscissa of the graph presented in Fig. 3-3 would be:

A. Point A
B. Point B

(A, *go to* 40)
(B, *go to* 41)

40 Point A shows that the highest frequency for any one value on the X axis was 6. The position of point A on the X axis shows that the value 3 occurred 6 times. Your answer was correct.

(*Go to* 42)

41 Point B is not the point that shows the highest frequency obtained by any value on the X axis. Point A shows a value or frequency of 6 on the Y axis, whereas point B has a frequency of 2. The Y axis indicates the frequency of occurrence for each and every value on the X axis. There the value of point A is 3 with a frequency of occurrence of 6. Your answer was not correct.

(*Go to* 42)

42 Frequency polygons are useful for comparing various groups with respect to a given characteristic. For instance, in Fig. 3-2, which was presented earlier, two age categories of children can be compared for their systolic blood pressure. From the frequency distribution plotted on the graph for each group (9- and 10-year-olds), one can visually judge whether systolic blood pressure differed radically between the two groups. Would you judge the two groups of children to have radically different blood pressures?

A. Yes
B. No

(A, *go to* 43)
(B, *go to* 44)

43 The path that each line follows is closely related to the other, so that most of the 10-year-olds in Fig. 3-2 have a systolic blood pressure of about 110 and most of the 9-year-olds are 105. Both begin and end at approximately the same reading, and both generally follow the same pattern of measurements with increases in values of systolic blood pressure. The two groups are not radically different.

The graph presented in Fig. 3-4 would better be judged as containing rather different groups. This figure, from data presented in Anders (1971), indicates that those students organizing the short-term memorization of concepts in groups of 4's rather than groups of 3's had a more consistent recall pattern. The group organizing the concepts in groups of 3's for recall purposes were able to memorize the seventh item well, but they were not able to recall other items with any degree of consistency.

In a later chapter you will be presented with specific facts and techniques to determine just how different two frequency distributions really are. You will be presented with information to determine whether they differ only because of random or chance variation or because of real differences.

(*Go to* 45)

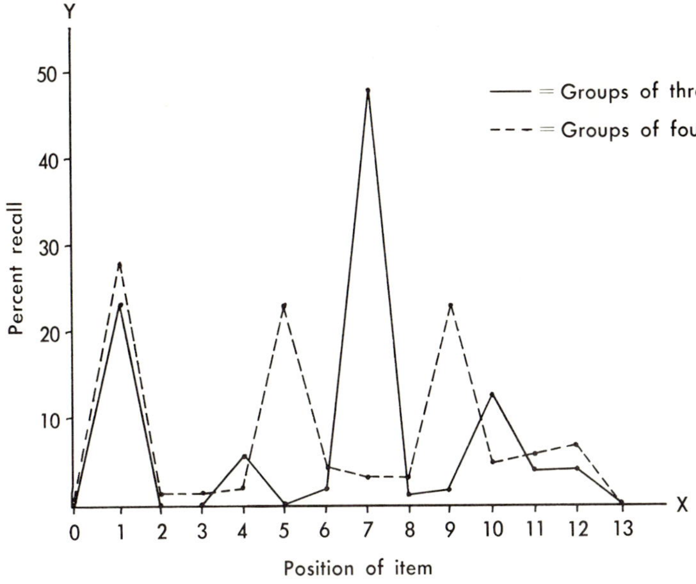

FIG. 3-4. Recall of items in short-term memory, by cognitive organization, groups of 3's and 4's. (Modified from Anders, T. R.: J. Exp. Psych. **90:**251-257, 1971.)

44 We agree with your judgment. The path that each line followed was closely related to the other, so that most of the 10-year-olds in Fig. 3-2 have a systolic blood pressure of about 110 and most of the 9-year-olds were at 105. Both began and ended at approximately the same reading and both generally followed the same pattern of number of measurements with increases in values of systolic blood pressure.

The graph in Fig. 3-4 would perhaps be better judged as containing rather different groups. This figure indicates that those students organizing the short-term memorization of concepts in groups of 4's rather than groups of 3's had a more consistent recall pattern. The group organizing the concepts in groups of 3's for recall purposes were able to memorize the seventh item well, but they were not able to recall other items with any degree of consistency.

In a later chapter you will be presented with specific facts and techniques to determine just how different two frequency distributions really are. You will be presented with information to determine whether they differ only because of random or chance variation or because of real differences.

(*Go to* 45)

---

45 When constructing a frequency polygon, it will often be necessary to use the interval scale such as the one constructed earlier for the frequency distribution. This scale is plotted on the X axis, since it represents the variable being measured. For any one of the intervals, one can make an assumption that if all the scores in that interval were averaged, the average would be equal to the midpoint of the interval. It then follows that the midpoint is the value that best represents any given interval. The interval scale on the graph in Fig. 3-5 is on:

A. The Y axis
B. The X axis

(A, *go to* 46)
(B, *go to* 47)

---

46 The frequency calibration is on the Y axis, not the interval scale. The X axis shows intervals of 5; e.g., 40-45, 60-65. Notice that for each point indicating a frequency for a certain interval, the point is located at the middle or midpoint of that interval.

(*Go to* 48)

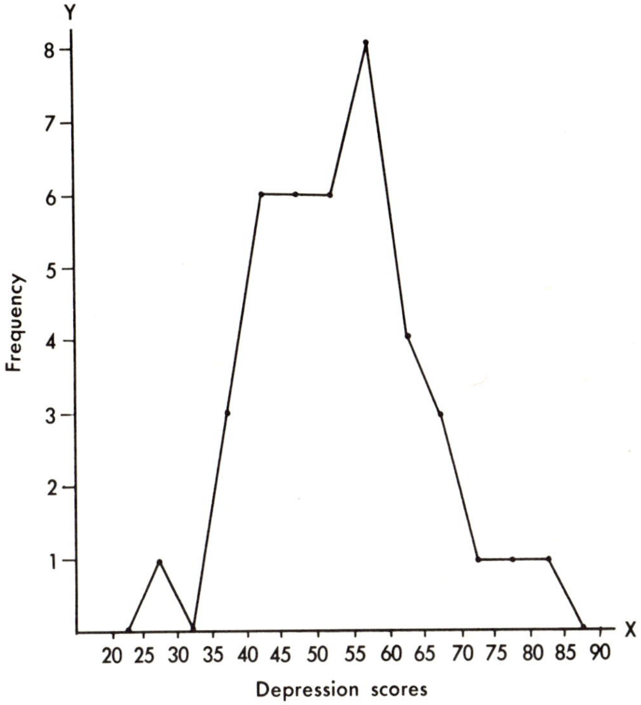

FIG. 3-5. Frequency polygon for depression scores.

---

47 The X axis indicates the interval scale with intervals of 5; e.g., 40-45; 60-65. The Y axis is the frequency calibration. Notice that for each point indicating a frequency for a certain interval, the point is located at the middle or midpoint of that interval. (*Go to* 48)

---

48 When dealing with midpoints of an interval, you should realize that you are not counting numbers between two points, i.e., end points of the intervals, but instead you are counting the number of *subintervals* that exist between these two end points. For example, between the interval limits of 10 and 20 there are ten subintervals (10 to 11; 11 to 12; 12 to 13; etc.). The midpoint falls in the middle of that interval.

Referring to Fig. 3-5, for the interval 85-90, the point of frequency for that interval is located:

A. At 87.5 on the X axis and at 1 on the Y axis (A, *go to* 49)
B. At 87 on the X axis and at 0 on the Y axis (B, *go to* 50)
C. At 87.5 on the X axis and at 0 on the Y axis (C, *go to* 51)

---

49 The frequency reading of the Y axis for the interval 85-90 should be zero. You are correct in saying that 87.5 is the point on the X axis because it is the midpoint for the 5-point interval of 85-90, since 87.5 is the middle of the subinterval 87 to 88. On any frequency polygon if there is no frequency greater than zero for any one interval on the scale, the point is plotted at zero on the Y axis.

(*Go to* 52)

50 The midpoint on the interval scale with 5 points would indicate that the plotting on the X axis should be 87.5, which is the number midway between 85 and 90. The answer 87.5 is the midpoint for the subinterval 87 to 88, which is the middle subinterval in the interval 85-90; i.e., 85-86=first subinterval, 86-87=second subinterval, and so on.

The part of the statement you chose that said the point is also plotted at zero on the Y axis is correct. Whenever an interval has registered no tabulation for that interval, it is recorded as zero on the frequency polygon.

(*Go to* 52)

51 The midpoint for the 5-point interval scale at 85-90 is 87.5, since it is the middle of the middle subinterval (87-88). Also, you are correct by saying that there was no tabulation of any phenomena for that interval. If no tabulation was registered for a given interval, it is plotted as zero on the Y axis.

(*Go to* 52)

52 An excellent example of the use of a frequency polygon can be seen in the work of de Charms and Carpenter (1968) who studied risk-taking among culturally disadvantaged students. They found that as the probability of success in a task increased from zero probability to 50% probability, students with high achievement scores (n Achievement) were more likely to choose the tasks with higher and higher probabilities of success. The increase in likelihood of choosing a task increased for these students until the probability of success reached 50%, and then the likelihood of choosing the task fell precipitously. The frequency polygon clearly demonstrates this pattern.

(*Go to* 53)

53 Another type of graph that is similar to the frequency polygon is the *histogram*. The construction of the histogram is the same as that of the frequency polygon except for the plotting of the frequencies. Fig. 3-6 shows a histogram constructed from the data given in Table 3-1 representing depression scores for mental patients.

Could you make a frequency polygon only on the basis of the information given in the histogram?

A. Yes

B. No

(A, *go to* 54)
(B, *go to* 55)

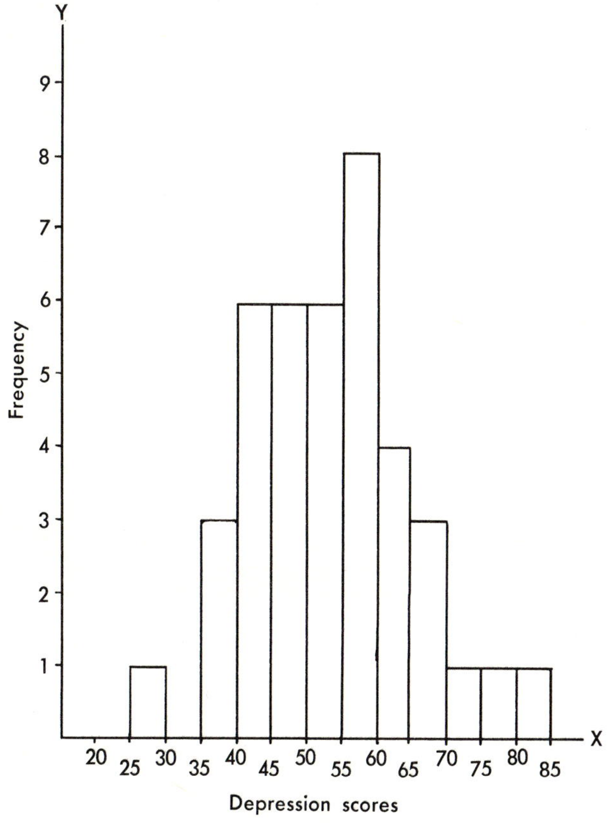

FIG. 3-6. Histogram of depression scores.

54 The scales on the X and Y axes are the same as they would be on the frequency polygon; therefore the histogram can easily be replaced by the frequency polygon. Since the frequencies are given for each interval, all that would be needed would be to plot a point at the midpoint of each interval at the spot where each bar ends on the frequency scale. After the points were connected, the result would be a frequency polygon.

(*Go to* 56)

55 Because the X and Y axes are the same on a frequency polygon as on a histogram, and since the frequencies are indicated for each interval, points can be plotted at the midpoint of each interval at the point of highest frequency. When the points are connected with straight lines, the result will be a frequency polygon.

(*Go to* 56)

56 Had you attempted to plot the data from Table 3-1 on a histogram such as the one found in Fig. 3-6, you would have had a problem in deciding what to do with the scores 65, 60, and 55; that is, would you have plotted the score 65 in the interval 60-65 or in the interval 65-70? To avoid this problem, it is necessary to establish *real limits* (some texts refer to these as true limits).

Real limits are theoretical boundaries that begin .5 unit below the stated lower limit and end .5 unit above the stated upper limit. For example, the bar for the interval beginning at 65 represents scores between 65 and 69; it therefore begins at the point 64.5 and ends at the point 69.5 on the X axis. In this way you would have been able to place the score 65 unambiguously in that interval.

The bar beginning at 40 and representing the interval 40-44 starts at 39.5 and ends at 44.5. The numbers 34.5, 39.5, 44.5, etc. are called real limits.

The histogram presented in Fig. 3-6 and in most of the literature uses a shorthand method of stating real limits. Rather than presenting 34.5-39.5, 39.5-44.5, etc. as interval limits, we simply use 35-39, 40-44, etc. and represent these on a histogram or frequency polygon with the beginning or lower limit of each interval.

The lower and upper *real limit* for the highest interval in Fig. 3-6 would be:

A. 79.5 to 84.5
B. 80 to 84
C. 80.5 to 85.5

(A, *go to* 57)
(B, *go to* 58)
(C, *go to* 59)

---

57 Just as 24.5 through 29.5 were the real limits for the bottom interval of 25-29, 79.5 through and including 84.5 are the real limits of the top interval, 80-84.

(*Go to* 60)

---

58 The limits for the interval in shorthand form are 80-84. The real limits would be 79.5 through and including 84.5.

(*Go to* 60)

---

59 The lower real limit is .5 less than the shorthand form of the lower limit (i.e., 80), and the upper real limit is .5 more than the shorthand form of the upper limit (i.e., 84). This makes the real limits 79.5 through and including 84.5.

(*Go to* 60)

---

60 The frequency polygon and histogram are used for essentially similar data. The choice of which to use is dependent on the desire of the individual preparing the figure. Both, however, demand interval data for the X axis (abscissa). It is important to distinguish between the histogram and the bar graph in that the bar graph uses data that are not intervally scaled and therefore is presented with the various bars separated.

You have now completed looking at the essential elements of statistical graphing. The three types of graphs are bar graphs, frequency polygons, and histograms. These three are representative of the type of graphs most frequently found in the scientific literature. The next part of this chapter will explore the area of statistical curves from distributions or arrays of data. It will also show where the smooth curve must be based on interval data. Specifically, the characteristics of *skewness* and *kurtosis* will be discussed.

After gaining some experience with frequency polygons and after you have read literature containing graphs like the frequency polygon, you will notice that sometimes the "peak" of some of the distributions will follow characteristic shapes. The "peak" refers to the points of highest frequency on the graph. Some graphs have only one peak, like the example of childhood blood pressure in Fig. 3-2, and some may have more than one peak per distribution.

Many times the peak will be at the extreme left or right end of the X axis. Where the peak is markedly displaced from the center of the X axis, we refer to the resulting curved as being *skewed*. If the peak is at the left

Continued.

side of the X axis, there is usually a tapering line or *tail* to the right of the peak. When the peak is to the left, it means that:

A. Measures were not frequent at the low or bottom end of the X axis scale
B. The highest frequencies occurred toward the low or bottom end of the X axis scale
C. The highest frequencies occurred toward the high or top end of the X axis scale

(A, *go to* 61)
(B, *go to* 62)
(C, *go to* 63)

---

61 If the distribution showed a peak to the left and a tail to the right, it means that the most frequent measures were those at the low end of the X axis scale. Your answer is not correct. The Y axis contained the high frequencies at the beginning of the X axis scale. A skewed curve with this characteristic of peak to the left and tail to the right is said to represent a *positively skewed* or *right-skewed* distribution. For the reverse situation, with tail to the left and peak to the right, the distribution is *negatively skewed* or *left-skewed*. The distribution in Fig. 3-7 represents income in the United States and would therefore be:

A. Negatively skewed
B. Positively skewed

(A, *go to* 64)
(B, *go to* 65)

---

62 If the distribution showed a peak to the left and a tail to the right, it indicated that the most frequent measures where those at the low end of the X axis scale, as you indicated. Under this condition the distribution is said to be *positively skewed* or *right-skewed*. When the tail extends to the left and the peak is at the right end of the X axis, the distribution is *negatively skewed* or *left-skewed*. The distribution in Fig. 3-7 represents income in the United States and would therefore be:

A. Negatively skewed
B. Positively skewed

(A, *go to* 64)
(B, *go to* 65)

---

63 The left end of an X axis is reserved for the lowest scores of the variable being measured. Therefore if the distribution showed a peak to the left and a tail to the right, it means that the most frequent measures were those at the low end of the X axis scale. Your answer is not correct. The Y axis contained the high frequencies at the beginning (left) of the X axis scale. A skewed curve with this characteristic of peak to the left and tail to the right is said to represent a *positively skewed* or *right-skewed* distribution. For the reverse situation, with tail to the left and peak to the right, the distribution is *negatively skewed* or *left-skewed*. The distribution in Fig. 3-7 represents income in the United States and would therefore be:

A. Negatively skewed
B. Positively skewed

(A, *go to* 64)
(B, *go to* 65)

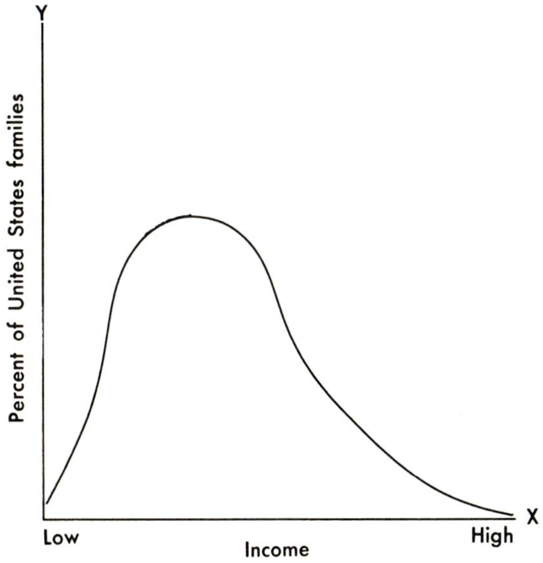

Fig. 3-7. Distribution of United States family income.

---

64 The peak is to the left and the tail is to the right. This indicates that the distribution is *positively* skewed and that the *least* frequent measures were those at the high end of the X axis. Your answer was not correct.

(*Go to* 66)

---

65 The distribution is positively skewed, as you indicated. The tail is to the right and the peak is to the left end of the X axis scale.

(*Go to* 66)

---

66 It is important to recognize that it is the *tail* that determines whether the distribution is skewed positively or negatively, not the pile-up of scores or measures (i.e., peak). If the tail stretches out to the left, the distribution is:

A. Symmetrical
B. Negatively skewed
C. Positively skewed

(A, *go to* 67)
(B, *go to* 68)
(C, *go to* 69)

---

67 One form of a symmetrical distribution is a curve in which both tails have equal frequencies, and the highest frequency of scores is in the center of the X axis scale. Fig. 3-8 represents a symmetrical distribution. Later the visual term "symmetrical" will be replaced by "normal" for this type of distribution.

A curve is said to be negatively skewed when the longer tail is to the left end of the X axis scale.

(*Go to* 70)

---

68 The tail to the left indicates that there were fewer measures at the low end of the X axis scale. It therefore means that, by definition, it is negatively skewed. When both tails have equal frequencies and the greatest frequency of scores is in the center of the X axis scale, as in Fig. 3-8, this curve is termed a symmetrical distribution. Later the visual term "symmetrical" will be replaced by "normal" for the type of distribution presented in Fig. 3-8.

(*Go to* 70)

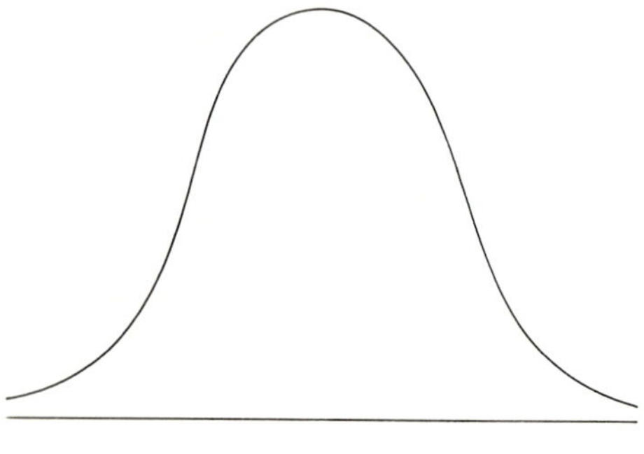

Fig. 3-8. Normal curve.

69 A curve is said to be negatively skewed when the tail is to the left end of the X axis scale; your answer is incorrect. When both tails have equal frequencies and the greatest frequency of scores is in the center of the X axis scale, as in Fig. 3-8, the curve is termed a symmetrical distribution. Later the visual term "symmetrical" will be replaced by "normal" for the type of distribution presented in Fig. 3-8.

(*Go to* 70)

---

70 Distribution curves can also be classified on the basis of how much peak (peakedness) they have rather than where the peak is located on the X axis. The relative extent of peakedness is called the *kurtosis,* which refers to the Y axis, of the distribution. *Skewness* is the term that refers to the distribution of scores along the X axis.

Fig. 3-9 shows three types of kurtosis. Curve 1 is called a *leptokurtic* curve; curve 2, a *mesokurtic* curve; and curve 3, a *platykurtic* curve.

Platykurtic (curve 3 in Fig. 3-9) is the term for a curve whose kurtosis is designed:

A. With a low peak
B. With a high peak

(A, *go to* 71)
(B, *go to* 72)

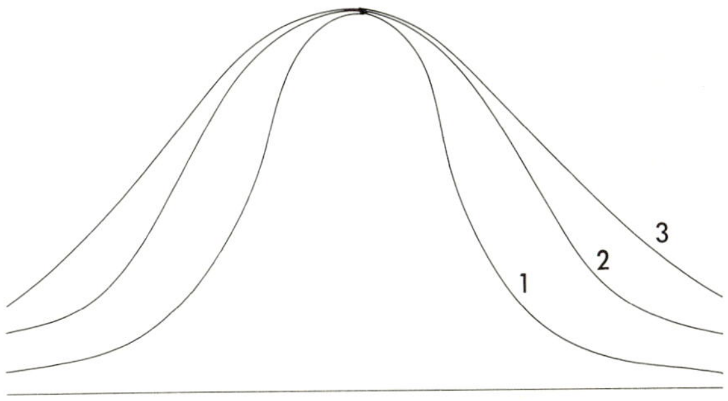

Fig. 3-9. Three types of kurtotic distributions: *1,* leptokurtic; *2,* mesokurtic; *3,* platykurtic.

---

71 The platykurtic curve shows a low peak and a broad middle section. The leptokurtic curve is the curve with the high peak and narrow middle (curve 1 in Fig. 3-9), and a middle level of peakedness is called a mesokurtic curve.

(*Go to* 73)

72 The leptokurtic curve is the type of kurtosis that shows a relatively high peak and narrow middle. Curve 1 in Fig. 3-9 is a leptokurtic curve. Curve 3 is a platykurtic curve and has a low level of peakedness and broad middle section; that is, the peak contains relatively few frequencies. A middle level of peakedness is called a mesokurtic curve.

(*Go to* 73)

73 As indicated by the three types of kurtosis shown in Fig. 3-9, kurtosis describes:

A. Height and breadth of the peak
B. Direction of skewness of a curve
C. Number of peaks

(A, *go to* 74)
(B, *go to* 75)
(C, *go to* 76)

74 Kurtosis describes the *height* of the peak and breadth of the middle section relative to other curves. Your answer is correct.

(*Go to* 77)

75 Skewness indicates a greater frequency of scores on one or the other end of the X axis; however, it does not indicate the height of the frequency of scores (peak). Your answer was incorrect. Kurtosis describes the height of the peak and breadth of the middle section relative to other curves.

(*Go to* 77)

76 The number of peaks is not indicated by the concept of kurtosis. You are incorrect. Kurtosis describes the height of the peak and breadth of the middle section relative to other curves.

(*Go to* 77)

77 If you recall from the section on frequency distributions, there was a column in the distribution table labeled the *cumulative proportion*. (See Table 3-3.) This was the proportion of the total number of measures recorded prior to the upper limit of a given interval on the X axis scale. If one arranges the interval scale of the measures along the X axis and an interval scale of from 0 to 1 on the Y axis, a curve can be plotted to graphically represent the cumulative proportions. (For cumulative percentages, the Y axis scale would range between 0% and 100%.) This kind of curve is called an *ogive curve* (pronounced ō′jīv).

The *ogive curve* is a graph describing:

A. Cumulative proportions
B. Cumulative frequencies
C. Frequency distributions

(A, *go to* 78)
(B, *go to* 79)
(C, *go to* 80)

---

78 The ogive curve describes the cumulative proportions (or cumulative percentages) that were derived from the cumulative frequency column in the distribution table.

(*Go to* 81)

---

79 The ogive curve does not represent a cumulative frequency distribution. It represents cumulative proportions (or cumulative percentages), which are derived from the values of the cumulative frequency column in the distribution table.

(*Go to* 81)

---

80 The ogive curve does not represent a frequency distribution. It represents the cumulative proportions (or cumulative percentages), which are derived from the frequency distribution table.

(*Go to* 81)

---

81 For the cumulative percentages, the scale is calibrated on the Y axis in an interval scale of 0-100 (instead of 0-1, as would be the case when using cumulative proportions). Fig. 3-10 is an ogive curve plotted for the cumulative percentages calculated in Table 3-3 representing depression scores for mental patients. Rather than plotting each interval, every other interval has been plotted. Notice that the points plotted on the X axis of the ogive curve are *not* at the midpoint of each interval, but are at:

A. Various places, depending on the interval
B. The upper limit of each interval

(A, *go to* 82)
(B, *go to* 83)

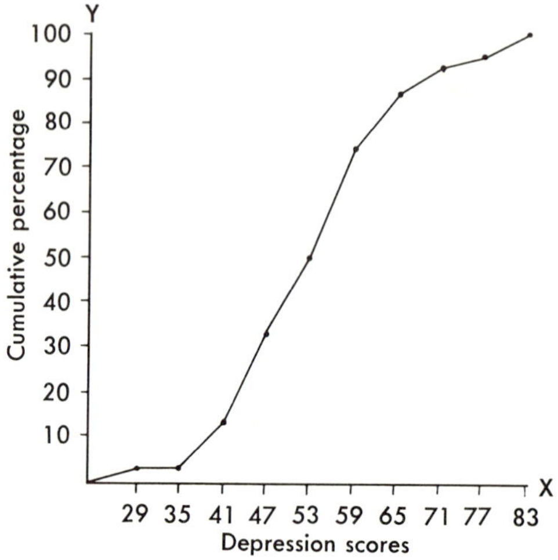

FIG. 3-10. Ogive curve for depression scores.

---

82 The points are not plotted at various points on the interval. If you look at the X axis scale in Fig. 3-10, you will notice that points are plotted at interval markings representing the *upper limits* of each interval. Therefore, if the interval is 42-44, the plotting point for that interval is 44. This is a rule for constructing all ogive curves, and it is unlike plotting a frequency polygon, which uses the midpoint of each X axis interval as the plotting point.

(*Go to* 84)

---

83 The rule in constructing the ogive curve is that each plotting point for the X axis is at the *upper limit* of each interval. If the interval is 42-44, the plotting point for that interval is 44. It is unlike the frequency polygon, which is plotted at the midpoint of each interval on the X axis.

(*Go to* 84)

---

84 What this ogive curve indicates is that the cumulative percentage of depression scores starts to accumulate rather slowly, begins to rise rapidly beginning at the interval with an upper limit of 41, and slows down again at the interval with an upper limit of 62. Ogive curves are highly useful for visually observing the cumulative distribution of scores (in percentages or proportions) in terms of how they change as they increase from low to high in value.

(*End of chapter* 3)

---

chapter

4

Measures of central tendency

At the conclusion of this chapter, you should be able to:

1. Apply the label "measures of central tendency" to the concept of some kind of central value for a distribution of data.
2. Name at least three measures of central tendency, including mode, median, and mean, and associate the terms correctly with a list of definitions.
3. Develop an understanding of the concepts *mode* (value having greatest frequency), *median* (center value), and *mean* (sum of all values divided by the number of values), and demonstrate this by calculating the mode, median, and mean for a given array of data.
4. Demonstrate that there can be more than one mode in a distribution by correctly drawing and naming an illustrative example of a two-mode ("bimodal") curve.
5. State which measures of central tendency are appropriate for each scale of measurement.
6. Calculate a sample mean, given a frequency table of grouped data, by using the interval midpoint for the score for all cases falling within the interval.
7. Compute and write down the deviations about the mean for a data array, and state what the sum of those deviations should be.
8. State the conditions under which the median is preferred to the mean as a measure of central tendency.
9. Demonstrate the effect of skewness on measures of central tendency by correctly designating the relative location of the mode, median, and mean on two curves, one symmetrical and one skewed.
10. State which measure of central tendency would usually show less variation (be more stable) over repeated sampling from a population.

11. Given an appropriate data table (N, $\overline{X}$, and Md) for three samples of unequal size:

    (a) compute an unweighted mean of means ($\frac{\Sigma\overline{X}_i}{j}$);

    (b) compute a weighted mean of means ($\frac{\Sigma N_i\overline{X}_i}{\Sigma N_i}$); and

    (c) indicate that the overall (weighted) median cannot be calculated from the data.

---

1 A statistic is a number that lets us reduce some of the information in a set of data down to a single value. A statistic should be meaningful and understandable or easily interpreted. One of the most useful types of statistics is that which describes a typical value or the center of the distribution of a set of data. Such a measure is called a *measure of central tendency*. This chapter will present three such measures: the mode, the median, and the mean.

(*Go to* 2)

---

2 The simplest measure of central tendency in a distribution is the value or category having the greatest frequency. This is called the *mode*. The mode (or modal value) is the score or value that occurs most often. In Fig. 4-1, what is the mode of the frequency distribution?

A. 50
B. 70
C. 80

(A, *go to* 3)
(B, *go to* 4)
(C, *go to* 5)

FIG. 4-1

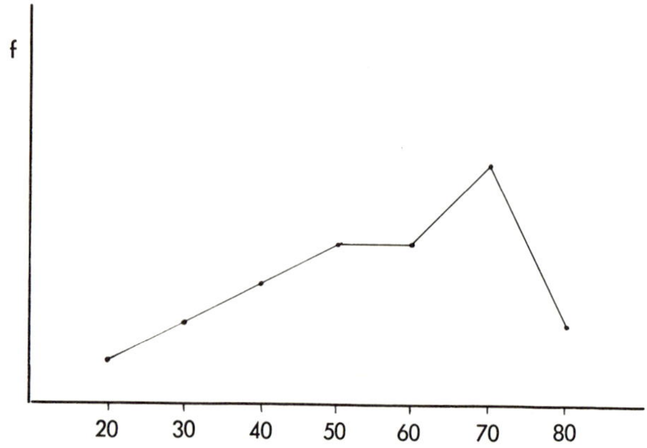

---

3 Your answer, 50, is the score or value occurring in the center of the range of scores. It is not the most common or frequently occurring score. The mode is the score having the greatest frequency; in this case, 70.

(*Go to* 6)

---

4 The mode is the score occurring most frequently. Your answer of 70 is correct.

(*Go to* 6)

---

5 Your answer, 80, is the highest score or value shown in the distribution. The mode is not the highest score, or score with the greatest value, but the score with the greatest frequency of occurrence. In Fig. 4-1 the mode is 70.

(*Go to* 6)

---

6 In a study of achievement motivation in delinquent boys, Tutt (1973) found that his sample from a home for delinquent boys showed less achievement identification than a matched control sample. Table 4-1 presents the frequency of occurrence for each score (a high score shows more achievement motivation).

What is the mode for the control group? After you have determined your answer . . .

(*Go to* 7)

TABLE 4-1. Frequency distribution of achievement motivation*

Score	*Delinquent group*	*Control group*
1	—	—
2	9	—
3	8	1
4	14	7
5	20	9
6	17	8
7	15	11
8	11	28
9	—	30

*Based on data from Tutt, N. S. 1973. Achievement motivation and delinquency. Br. J. Soc. Clin. Psych. **12**:225-230.

7 The mode for the control group in Table 4-1 is 9; that is, a score of 9 on achievement motivation occurred with the greatest frequency. Note that the mode is not 30, which is the number of times a score of 9 occurred in the control group. The mode is the value of the variable of interest (achievement motivation in Table 4-1) that occurs most frequently. Note also that the mode in this example is nowhere near the center of the range of scores in the sample.

Another study tabulated the kinds of stresses and psychologic reactions of nurses working in a disaster area after a hurricane (Laube, 1973). Table 4-2 presents frequencies for the different stress categories reported.

What is the mode (modal category) for Table 4-2? After you have determined your answer . . .

(*Go to* 8)

TABLE 4-2. Frequencies of major stress categories in nurse study*

1. Excessive physical demands	8
2. Concern for own/patient safety	8
3. Inadequate supplies	4
4. Concern for own family	2
5. Seeing the children	2
6. Increased medical responsibility	1
7. Seeing the poor suffer	1
8. Disorganization	1

*Based on data from Laube, J. 1973. Psychological reactions of nurses in disaster. Nurs. Res. **22:**343-346.

8 The modes for the stess categories in Table 4-2 are categories 1 and 2: excessive physical demands and concern for safety. Two points should be noted. First, more than one mode can exist in a distribution. In this case there were two modes. In the frequency distribution in Fig. 4-2, two modes also exist. In the curve illustrated, the two modes are not identical in frequency, but are clearly two high points in the curve and thus should be treated as two modes.

When you have more than one mode, the distribution is called *multimodal.* In Fig. 4-2 and in Table 4-2, since there are two modes, the distributions are called *bimodal.*

The second point to note is that the mode can be used with any level of measurement: nominal, ordinal, or interval. Table 4-2 involves a nominal level of measurement (categories of stress). In Table 4-2 the categories could be rearranged in any order of sequence, since order has no meaning in a nominal variable. In any order, the modes or most frequent occurrences would still be the categories "physical demand" and "concern for safety." For variables at the ordinal or interval level, there are measures of central tendency more useful than the mode.

(*Go to* 9)

FIG. 4-2

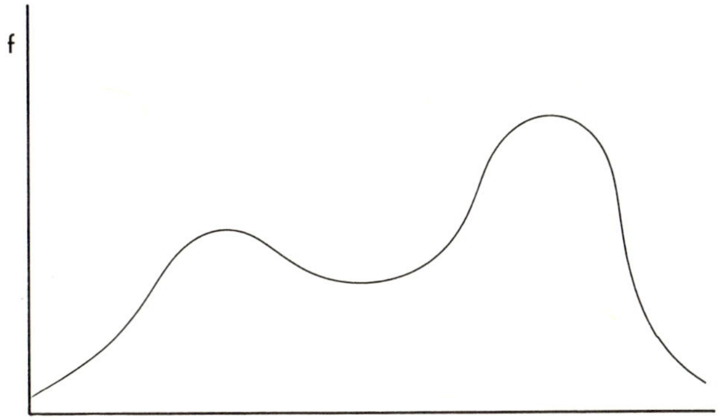

---

9 The *median* is that score in an ordered distribution above which and below which half the frequencies fall. It is the middle score in an ordered set of scores. In terms of percentiles, which will be discussed more in chapter 5, the median is the score at the 50th percentile. In the following distribution: 27, 32, 45, 49, 51—the median is 45, the middle score.

Using the data in Table 4-1, find the median for both the delinquent group and the control group. After you have determined your answer . . .

(*Go to* 10)

---

10 Since there are 94 cases in each sample in Table 4-1, the median is the score held by the middle individual; i.e., the 47th and 48th person. Thus an achievement motivation score of 5 is the median for the delinquent group, and a score of 8 is the median for the control group. These statistics were found by cumulating frequencies until the 47th and 48th persons were found.

In some distributions, the median score may not exist in the scores actually obtained. For example, look at this distribution: 53, 62, 68, 74, 77, 82.

The median must lie between ____________ and ____________.

(*Go to* 11)

---

11 The median must lie between 68 and 74.

53, 62, 68	74, 77, 82
3 scores	3 scores

In this case the median is defined as the arithmetic mean of the middle scores. Thus the median is $\frac{68 + 74}{2} = 71$.

In some cases the variables are lumped together into intervals (e.g., ages 20 to 25). In these cases, although the median interval can be determined, the actual value of the median within the interval is somewhat more complicated to calculate. In the nursing stress study referred to earlier (Laube, 1973), the data in Table 4-3 were shown. The median interval for all nurses in Table 4-3 was 17-24 hours. That is where the 14th nurse fell in the distribution (since with 27 nurses, the score for the 14th nurse was the middle score). A more precise median is calculated as follows. Remember from chapter 3 that an interval is assumed to have real limits 0.5 above and below its integer boundaries. Thus the real limits of the interval 17-24 are 16.5 and 24.5. There are 11 cases in the first two categories, and 9 more in the third category. Since we want the 14th case, we need to go 3 more cases or 3/9 of the way into the third category or interval to reach the 14th case. That is, we want to go 3/9 of the way into the 8-hour interval of $16.5 - 24.5$. The median, then, is calculated as $16.5 + 3/9 \times 8 = 16.5 + 2.67 = 19.17$ hours.

The calculation of positional measures within an interval is dealt with at more length in chapter 5.

(*Go to* 12)

TABLE 4-3. Hours of continuous work experienced by nurses in disaster study*

Hours	*Frequency*
0-8	3
9-16	8
17-24	9
25-32	6
33-40	0
41-48	1

*Based on data from Laube, J. 1973. Psychological reactions of nurses in disaster. Nurs. Res. **22**:343-346.

12 A study was made of the causes of 299 consecutive deaths occurring in an intensive care/special care nursery for newborns with major problems (Duff and Campbell, 1973). The causes are indicated in Table 4-4. What is the median problem for the data in Table 4-4?

(*Go to* 13)

TABLE 4-4. Major problems associated with infant death in a special care nursery*

Problem	*Number of deaths*
Respiratory	108
Extreme prematurity	60
Heart disease	42
Multiple anomalies	29
Trisomy	8
Meningomyelocele	7
All other	45
	299

* Based on data from Duff, R. S., and A. G. M. Campbell. 1973. Moral and ethical dilemmas in the special-care nursery. New Eng. J. Med. **289:**890-894.

13 Indeed, there can be no median for the data in Table 4-4 because there is no order to the data and therefore we cannot speak of a middle. Although the question was somewhat of a trick question, the issue is real: the statistic used must be appropriate for the data being analyzed. Another characteristic of the median can be observed by noting the median of the following two distributions:

21, 31, 36, 47, 52   21, 31, 36, 47, 98

The medians for both distributions are 36. This is true even though one distribution includes a maximum score of 52 and the other a maximum score of 98. Noting the results in the preceding example, would you say the median is sensitive or insensitive to extreme scores?

A. Sensitive
B. Insensitive

(A, *go to* 14)
(B, *go to* 15)

14 To be sensitive to extreme scores means that the introduction of an extreme score (or the making more extreme of the previous maximum or minimum score) would produce notable changes in the value of the median. To be insensitive to extreme scores means that changes in the value of the scores at the ends of the distribution do not change the value of the median or middle of the distribution.

(*Go to* 15)

15 As the two cases in frame 13 show, the median is insensitive to, and, in fact, unaffected by extreme scores. All of the scores above the median could double their value, and the value of the median would still remain the same. Introducing an additional score, however, will move the median slightly toward the value of the new score. You might note that the mode is totally unaffected by extreme scores (unless the frequency of any one score becomes greater than the former mode).

To use the median as a measure of central tendency, what is the lowest level of measurement that the data can have?

A. Nominal
B. Ordinal
C. Interval

(A, *go to* 16)
(B, *go to* 18)
(C, *go to* 17)

16 It is impossible to obtain a median on nominal data, since one of the prerequisites for calculating the median is rank-ordering of the data. Nominal data, which deal with categories such as race and sex, cannot be rank-ordered. Your answer, nominal, is incorrect; you must have at least ordinal data.

(*Go to* 19)

17 Although you can use the median with an interval level of measurement, it is not the lowest level of measurement on which the median can be used. The median requires that the data be rank-ordered, and the lowest level of measurement for which data can be ordered is on the ordinal level. By using the median with interval data you are not using the measure of central tendency most appropriate to this level of measurement.

(*Go to* 19)

---

18 It is necessary to rank-order data before the median can be calculated. Since the essence of ordinal data is the fact that they can be rank-ordered, the ordinal level of measure is indeed, as you maintained, the lowest level of measurement possible for use of the median.

(*Go to* 19)

---

19 The most complex measure of central tendency, yet the best known, is the arithmetic *mean.* Nearly everyone is familiar with the process of summing a set of scores and dividing by the number of scores to obtain the mean. Sometimes the mean is referred to as the "average," but it is recommended that you not use the term "average" because of its imprecise meaning. People speak of the "average person," but they do not mean the "mean" person. "Average" also is often used to mean "within normal limits," which, again, is not the same as the mean.

The calculation of the mean requires that the data be at least interval in nature, since the process of adding or summing must take place. Adding requires numbers with equal intervals between place value, which is defined as an interval level of measurement. It is inappropriate to perform the arithmetic operations involved in finding the mean on data such as type of personality (nominal) or perceived severity of disease (ordinal).

Calculate the mean of the following scores: 2, 3, 5, 2. Mean = ______________.

(*Go to* 20)

---

20 The sum of the scores is 12. The number of scores is 4. The mean is the sum (12) divided by the number of scores (4) = 3.

The formula for the mean is stated as follows:

$$\overline{X} = \frac{X_1 + X_2 + X_3 + X_4 + \ldots + X_n}{N}$$

This equation may look forbidding, yet it simply states that the mean ($\overline{X}$, which can be pronounced as "X bar") equals the sum of the individual scores (X_1, X_2, X_3, etc.) divided by the number of scores (N).

The last figure in the above equation (X_n) is a statistical symbol representing the last possible entry in the series of numbers, which, when the formula is used with data, is replaced with actual data. Any symbol, usually in a capital letter with a bar drawn above ($\overline{X}$, $\overline{Y}$), represents the mean.

The equation for the mean can be more simply stated in notational form using the symbol for "the sum of" (Σ). Σ is the Greek letter sigma.

$$\overline{X} = \frac{X_1 + X_2 + X_3 + X_4 + \ldots + X_n}{N} = \frac{\Sigma X}{N}$$

Continued.

Again, the mean equals the sum of all the scores divided by the number of scores.

Table 4-5 shows a completed example of the calculation of the mean from a set of data in which the raw data have been organized to show the frequency of occurrence of each raw score.

(*Go to* 21)

TABLE 4-5. Calculating the mean from an ungrouped frequency distribution

X	f	fX	
25	1	25	$\bar{X} = \frac{\Sigma fX}{N}$
24	2	48	
22	5	110	$\bar{X} = \frac{303}{15}$
18	3	54	
17	3	51	$\bar{X} = 20.2$
15	1	15	
	$N = 15$	$\Sigma fX = 303$	

TABLE 4-6. Calculating the mean achievement motivation*

Score	Delinquent group		Control group	
X	f	fX	f	fX
1	—		—	
2	9		—	
3	8		1	
4	14		7	
5	20		9	
6	17		8	
7	15		11	
8	11		28	
9	—		30	
	$N =$	$\Sigma fX =$	$N =$	$\Sigma fX =$

$\bar{X}_{delinq} = \frac{\Sigma fX}{N}$ $\quad$ $\bar{X}_{control} = \frac{\Sigma fX}{N}$

$\bar{X}_{delinq} =$ $\quad$ $\bar{X}_{control} =$

*Based on data from Tutt, N. S. 1973. Achievement motivation and delinquency. Br. J. Soc. Clin. Psych. **12:**225-230.

21 Table 4-6 presents again the data from the Tutt (1973) study shown in Table 4-1. Table 4-6 provides the columns and steps for calculating the mean. Calculate the mean for both the delinquent group and the control group. When you are finished, to check your work . . .

(*Go to* 22)

22 Your calculations should look like the following:

Delinquent group

N = 94 ΣfX = 493

$$\bar{X}_{delinq} = \frac{\Sigma fX}{N} = \frac{493}{94} = 5.24$$

Control group

N = 94 ΣfX = 695

$$\bar{X}_{control} = \frac{\Sigma fX}{N} = \frac{695}{94} = 7.39$$

These calculations indicate that the control group had a higher achievement motivation score than did the group of delinquents.

Table 4-7 is a partially completed table of a grouped frequency distribution. These data are from the Laube (1973) nursing stress study. In determining fX for data in an interval, it is assumed that all the scores in a given interval lie at the midpoint of the interval. Thus the midpoint of the first interval in Table 4-7 is 4 hours, and the midpoint for the second interval is 12.5 hours. All three persons in the first interval are assumed to have worked 4 hours.

Complete Table 4-7 and calculate the mean. When you are done . . .

(*Go to* 23)

TABLE 4-7. Hours of continuous work experienced by nurses in disaster study*

Hours	*Midpoint* X	*Frequency* *f*	*f*X
0-8		3	12
9-16	12.5	8	
17-24		9	
25-32		6	
33-40		0	
41-48		1	
		N =	ΣfX =

$$\bar{X} = \frac{\Sigma fX}{N}$$

$\bar{X}$ = ______

*Based on data from Laube, J. 1973. Psychological reactions of nurses in disaster. Nurs. Res. **22:**343-346.

23 The completed Table 4-7 should look like this:

Hours	Midpoint X	Frequency f	fX	
0-8	4	3	12.0	$\bar{X} = \frac{\Sigma fX}{N}$
9-16	12.5	8	100.0	
17-24	20.5	9	184.5	$= \frac{512}{27}$
25-32	28.5	6	171.0	
33-40	36.5	0	0	$= 18.96$
41-48	44.5	1	44.5	
		N = 27	ΣfX = 512.0	

Or, the mean number of continuous hours work by the nurses was 18.96.

If you think that you are having trouble with the rationale behind these calculations for grouped data . . . (Go to 24)

Otherwise . . . (Go to 25)

24 Perhaps the problem with your calculation of the mean from these grouped data, not counting arithmetic errors, stems from not realizing that you are in reality "weighting" the various scores. The score is said to be the "midpoint" of an interval, and the "weighting" is the frequency of scores falling in that interval. When a student calculates a grade-point average, the same sort of "weighting" procedure is performed. An A is worth 4 points; a B, 3 points; and so on. You multiply the number of credit hours of B's received in a term by the "weight" of a B (where the number of credit hours = f, and the weight of the grade = X), and then divide by the number of credit hours (N).

(Go to 25)

25 Before moving on to discussing certain mathematical properties of the mean, a review of appropriate levels of measurement may be useful.

The mean is suited for use only on variables with (1) ____________ values or level of measurement. It cannot be applied to either (2) ____________ or ____________ values.

Choose the answer which correctly fills blanks (1) and (2).

A. (1) Ordinal, (2) nominal or interval
B. (1) Interval, (2) nominal or ordinal

(A, go to 26)
(B, go to 27)

---

26 The arithmetic mean should not be applied to nominal or ordinal values. The mean should be applied only to interval scales of measurement. In other words, you cannot find the mean for categories or for the order in which items may be placed.

(*Go to* 27)

---

27 The mean should be used with variables with interval values or levels of measurement. The mean calls for a quantification of values, not just a categorization (nominal) or ordering (ordinal) of items. One must be clear on the variable being measured to decide the measure of central tendency to use. Thus although you cannot find the mean handedness in a classroom (where the students are categorized as left-handed, right-handed, or ambidextrous), you could find the mean number of left-handed students among several classrooms.

(*Go to* 28)

---

28 Since the mean will be used as a basis for teaching measures of dispersion, it is important to understand the principle of deviations about the mean. One of the most important properties of the mean is that it is the point in a distribution about which the sum of all the deviations from the mean of the distribution equals zero. Take the scores 6, 7, 8, 8, 9, 10, for example. The mean for this distribution is 8; 7 is one less than the mean (deviation = −1); 6 is two less (d = −2); 9 is one more (d = +1); and 10 is two more (d = +2). Then (−2) + (−1) + (0) + (0) + (+1) + (+2) = 0.

In other words, $\Sigma(X - \overline{X}) = 0$, where X = individual score and $\overline{X}$ = mean.

One might compare the mean to the balance point or fulcrum of a board. This board can be in balance on a fulcrum with a small weight far from the fulcrum balancing a large weight near the fulcrum. This concept suggests a second property of the mean: the mean is sensitive to extreme scores when these scores are not balanced on both sides of the mean.

For example, calculate and compare the means of these two groups of data:

(1) 5, 8, 8, 11 $\overline{X}_1$ = ____________

(2) 5, 8, 8, 19 $\overline{X}_2$ = ____________

When you are done . . .

(*Go to* 29)

---

29 The substitution of the score of 19 for the score 11 seriously affects the mean value (shifting the mean from 8 to 10). The characteristic of the mean that it is affected by extreme values will be important when attempting to make decisions about which measure of central tendency to use on a set of data.

Now calculate and compare the means of these four groups:

(1) 7, 10, 12, 13 $\overline{X}_1$ = ________

(2) 7, 10, 12, 19 $\overline{X}_2$ = ________

(3) 1, 10, 12, 13 $\overline{X}_3$ = ________

(4) 2, 10, 12, 18 $\overline{X}_4$ = ________

When you are done . . .

(*Go to* 30)

---

30 The comparison of the four means should indicate something about how the mean changes with extreme scores:

$\overline{X}_1 = 10.5$ $\overline{X}_2 = 12$ $\overline{X}_3 = 9$ $\overline{X}_4 = 10.5$

Which of the following statements about the effect of extreme values on the mean is most true?

A. We cannot really tell the effect of extreme scores on the mean without calculating in each case.

B. We can tell, in general, what effect extreme scores will have on the mean.

(A, *go to* 31)
(B, *go to* 32)

---

31 Although it is true that the exact numerical effect an extreme score will have on a mean cannot be known without calculation, it is not true that we do not know the effect extremes will have on a mean. The general knowledge of the effect of extremes on the mean and other measures of central tendency is essential for quantitative reasoning in daily science or applied fields.

(*Go to* 32)

---

32 As the example illustrates, the extreme high score in group 2 pulled the mean toward the high end, the extreme low score in group 3 moved the mean toward the low end of the scale, and the two extreme scores (high and low) in group 4 resulted in no change in the mean. The mean always shifts toward the extreme score, whether the extreme is high or low, unless the extreme values are balanced exactly.

Next, consider briefly the problem of finding the mean of a group of means. A first-year medical student received first-semester grades of biochemistry 78, anatomy 89, and human behavior 84. These grades are summaries or means of all his semester work. The mean grade for the three courses would be a mean of means, $\bar{\bar{X}} = \frac{78 + 89 + 84}{3}$ $= 83.67$. This was a straightforward computation of means; a calculation of this type assumes that all the grades were equivalent in value. If, however, the biochemistry class was rated as a 5-hour course, anatomy as an 8-hour course, and human behavior as a 6-hour course, then what would be the mean grade for the three courses?

Calculate the mean, then . . .

(*Go to* 33)

---

33 To calculate the mean of the means for the three courses, we must use a *weighted mean* based on the relative worth of each course measured by its credit in semester hours:

Anatomy	$\bar{X} = 89$	89×8 sem hr =	712
Biochemistry	$\bar{X} = 78$	78×5 sem hr =	390
Human behavior	$\bar{X} = 84$	84×6 sem hr =	504
		19 sem hr =	1606

The mean of the means is then:

$$\bar{\bar{X}} = \frac{1606}{19} = 84.5$$

The same reasoning applies whenever you have different numbers of cases that enter into different means. For instance, three sections of a senior educational psychology class produced the following section means:

Section 1, 35 students, $\bar{X} = 84.2$
Section 2, 48 students, $\bar{X} = 82.4$
Section 3, 30 students, $\bar{X} = 78.5$

The mean of means is found by:

$$\frac{35 \times 84.2 + 48 \times 82.4 + 30 \times 78.5}{113 \text{ students}} = \frac{9257.2}{113} = 81.9$$

The algebraic formula to express computing the weighted mean of means is

$$\bar{\bar{X}} = \frac{\Sigma N_i \bar{X}_i}{\Sigma N_i}$$

where N_i is the number of cases or persons in set i or sample i (here, section 1, 2, or 3), and ΣN_i is the sum of all cases in all sets or samples.

(*Go to* 34)

---

34 You should recognize, from the earlier discussion on medians, that it is not possible to calculate an overall median for several samples or sets of data knowing only each sample median. For instance, what is the combined median for the two sets of data below?

Sample 1: 8, 10, 25
Median = 10

Sample 2: 7, 14, 14, 20, 30
Median = 14

(*Go to* 35)

35 If you take a value halfway between the two medians (12) you are, in effect, taking an unweighted mean of the two medians. Even taking a weighted computation of the two samples, knowing sample size and median, would give you a misleading figure (3 cases × 10 + 5 cases × 14 = 100/8 = 12.5). The correct overall median is the middle value of the two samples combined:

7, 8, 10, 14, 14, 20, 25, 30 Median = 14

Knowing only the sample medians and the number of cases in each sample does not give you enough information to calculate the overall median.

(*Go to* 36)

36 You have now been presented with the three most frequently used measures of central tendency: the mean, the median, and the mode. The mean is the most frequently encountered of the measures of central tendency. It also is the most stable measure; that is, if repeated samples were drawn from a population, the mean would vary less from sample to sample, as a rule, than would the median or mode.

It should be realized that in an exactly symmetric distribution, the mean, median and mode are at the same point or identical in value. Thus, in the normal distribution, mean = median = mode. The normal distribution, sometimes referred to as "bell-shaped," is a distribution that is symmetric and has known characteristics (that is, there is an equation that defines the normal distribution). Many characteristics of interest in the world turn out to be distributed approximately as the normal distribution. Because of this, the normal distribution will enter further into our discussion of statistics later in the book.

(*Go to* 37)

37 Recall the discussion of the effect of extreme values on each of the measures of central tendency. The mean is most affected and the mode least affected. In an otherwise normal distribution with some extreme values, the mean will be skewed toward the extreme values more than the median or mode. Samples whose distributions are not symmetric are sometimes best described by other than the mean. That is, the median may give a more fair picture of where the center is than does the mean. Whoever is doing the describing has to make that choice.

Tables 4-1 and 4-6 presented the data from the study of achievement motivation in delinquent youth (Tutt, 1973). You calculated the mean and median for the two samples in the study. Table 4-8 presents the data again along with the mode, median, and mean for each of the two samples.

Note the differences between the measures of central tendency for the mode, median, and mean. The delinquent group, which is more symmetric or more normal in distribution, has central tendency estimates close together, whereas the control group, which is highly skewed with many scores piled up at one end, has more disparate estimates of central tendency. The mean in the delinquent group is pulled only slightly toward the right or high scores, meaning that the curve is skewed to the right. The mean in the control group is pulled toward the left or low scores, meaning that the distribution in the control group is skewed to the left with most of the scores bunched at the higher end of the achievement motivation scale. The median score of 8 in the control group is probably more reflective of central tendency in the control group distribution than is the mean score of 7.4.

(Go to 38)

TABLE 4-8. Frequency distribution of achievement motivation*

Score		*Delinquent group*	*Control group*	
1		—	—	
2		9	—	
3		8	1	
4		14	7	
5		20	9	
6		17	8	
7		15	11	
8		11	28	
9		—	30	
		94	94	
				Difference
	Mode	5	9	4
	Median	5	8	3
	Mean	5.2	7.4	2.2

*Based on data from Tutt, N. S. 1973. Achievement motivation and delinquency. Br. J. Soc. Clin. Psych. **12:**225-230.

38 Because of the assumptions on which each measure of central tendency is based, they have different uses when dealing with statistics.

The *mode* assumes no more than a *nominal* level of measurement. Because the mode demands only this level of measurement, it can also be used with higher levels of measurement as well.

The *median* assumes at least an *ordinal* level of measurement. Therefore when the data that you are dealing with are ordinal, you can use both the median and the mode as the measure of central tendency.

Your choice of the median or mode should depend primarily on the extent to which you can afford to lose information in presenting a measure of central tendency. That is, the median uses much more information than does the mode when you calculate it, and therefore it is a more stable and sensitive measure than is the mode.

Furthermore, there are certain tests of statistical significance that use the median, whereas there are fewer tests that employ the mode. You will be introduced to these tests in later chapters.

The *mean* assumes that the level of measurement is *at least interval.* It would be impossible to use the mean dealing with an ordinal or nominal level of measurement, although you can apply the median and mode to a variable having an interval level of measurement. In so doing, however, you are losing information, since the mean uses the greatest amount of information of all measures of central tendency and is, therefore, most stable.

Whenever a distribution of interval data is highly skewed, it is wiser to use the median as the measure of central tendency, since the median is less affected by extreme scores than is the mean. However, most tests of statistical significance use the mean as the basis.

You should constantly keep in mind the fact that you can always use measures of central tendency that are appropriate for one level at any higher levels of measurement. For example, it is appropriate to use the median as a measure of central tendency when dealing with blood pressure (an interval variable).

However, you should not use measures that are appropriate at a higher level for lower levels of measurement; that is, it is inappropriate to use the median as a measure of central tendency on a nominal-level variable such as sex.

(*End of chapter* 4)

chapter
5
Measures of dispersion

At the conclusion of this chapter, you should be able to:

1. Apply the label "measures of dispersion" to the concept of measures that indicate the extent to which scores deviate from some measure of central tendency.
2. Name at least two measures of dispersion including standard deviation and either range or semi-interquartile range, and associate the names correctly with a list of definitions.
3. State the range of a given table of grouped data.
4. Given cumulative frequency distribution tables for ungrouped and grouped data, calculate the exact location in the distributions for any specified percentile.
5. Given a cumulative frequency distribution table with percentiles, indicate the quartile points in the distribution, and compute the semi-interquartile range.
6. Write correct interpretive statements about a data table involving range, percentiles, quartiles, and semi-interquartile range.
7. State whether the range or semi-interquartile range will (a) be distorted more by an extreme score and (b) show more consistency (stability) over repeated samples from some population.
8. Given two normal curves with different shapes (different standard deviations), indicate that the proportion of cases (area under the curve) between the points ± 1 standard deviation away from the mean will be the same in both curves.
9. State from memory that given a normal distribution, about two thirds of the cases will fall within 1 standard deviation of the mean and just over 95% of the cases will fall within 2 standard deviations of the mean.
10. Given an array of data and the formula for standard deviation, calculate s.
11. Given the data about two samples of equal size with equal means but different standard deviations, write correct interpretive statements

about the differences in the samples in terms of dispersion of scores about the mean.
12. State which measures of dispersion are appropriate for each scale of measurement.
13. Correctly pair the following symbols with a list of terms: N or n, $\overline{X}$, Σ, s, σ, f, %, Q.

---

1 The purpose of this chapter is to introduce various measures of dispersion. By dispersion we mean a measure that indicates the extent to which a set of scores deviate from, or disperse themselves around, some measure of central tendency for that set. A measure of dispersion is a descriptive statistic that, in a single value, tells you something about the spread of a set of scores. To fully understand this chapter, it is necessary that you have a good grasp of ideas dealing with distributions and measures of central tendency discussed in previous chapters.

Once the mean of a distribution has been determined, calculating the relationship of any given score in that distribution to the mean is a relatively simple arithmetical procedure. Consider the two frequency polygons showing the distribution of heart rate for two populations as depicted in Fig. 5-1. In which curve does a heart rate of 90 represent the greater deviation from the mean when compared with other scores in the distribution?

(A, *go to* 2)
(B, *go to* 3)

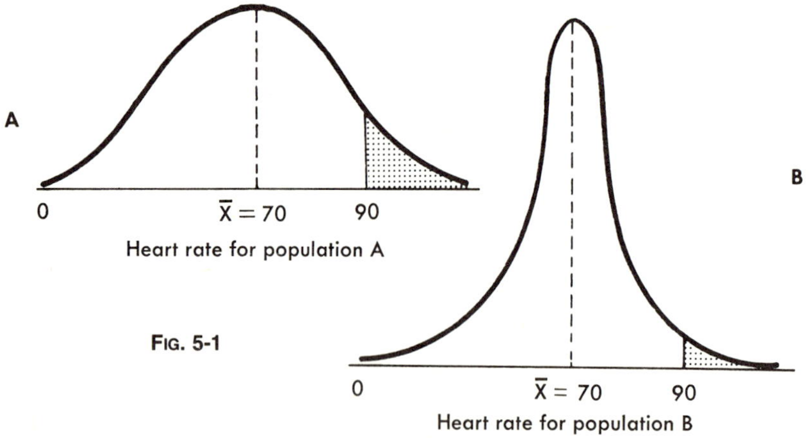

FIG. 5-1

---

2 Compare the shaded areas of the two curves. Although in both curves a score of 90 is 20 units away from the mean, when *compared with other scores* in the distribution, the two distributions are different. In curve A, several scores exceed 90, and a number of scores occurred at 90. In distribution B, 90 is one of the least frequent scores, and there are fewer scores above 90 than in curve A. Thus in distribution B a score of 90 heart beats a minute represents a more rare event or a greater deviation from the mean compared to other scores.

(*Go to* 4)

---

3 In B, only a small part of the total area under the curve falls above 90. A person with a heart rate of 90 had one of the highest scores of all persons in that distribution. B is therefore the correct answer because in that distribution a score of 90 beats a minute represents a more rare event or a greater deviation from the mean of 70 when compared to other scores.

(*Go to* 4)

---

4 From this one example it can be inferred that the kurtosis (peakedness) of a distribution has an effect on the interpretation we give to the relative distance of a given score from the mean. We can get a more accurate picture of the relationship between any individual score and the mean if we have some information about the *dispersion* of scores. Dispersion is therefore a useful concept for interpreting data. Together with measures of central tendency, measures of dispersion can tell us something about what kind of numerical values to expect in a distribution or array of data and how extreme any given value is. For example, a teacher may be interested in what kind of spread (i.e., distribution) of abilities students show in reading; a psychologist may be interested in how extreme a set of behaviors are from "normal" behavior; or a physician may be concerned with the extent to which a patient's vital signs deviate from the expected norm.

One crude but useful measure of dispersion of scores in a distribution is the *range*. The range is simply a statement of the difference between the smallest and largest scores in a distribution.

What is the range in Table 5-1?

(*Go to* 5)

TABLE 5-1. Ages of head of household in a sample of commodity food recipients

27	24	33	30	29
36	37	38	34	35
51	23	28	39	32
21	43	40	41	30
49	39	34	37	37

5 The range is the distance between the smallest and the largest scores. If you correctly scanned the scores you would find the youngest age to be 21, the oldest age 51. The range can be expressed as 21 to 51, or 30 years. (Note: In some instances the range is expressed as "high minus low plus 1" to be inclusive of all values in the series; thus the range of 21 to 51 would be called 31; when confusion may arise, state the actual high and low values.)

If scores are grouped into intervals, the range is indicated as the distance from the midpoint of the lowest interval to the midpoint of the highest interval, since the midpoints are used to represent all scores grouped into that interval. For example, in a distribution in which the lowest interval containing scores is 20 to 24 and the highest interval containing scores is 50 to 54, the range is 22-52, or the midpoints of the minimum and maximum intervals.

Choose the statement below that provides you with the most meaningful information for working with a child exhibiting difficulty in learning to read.

A. The child has a reading comprehension standard score of 35.
B. The child has a reading comprehension standard score of 35. The mean score for all students in the school is 55.
C. The child has a reading comprehension standard score of 35. The mean score for all students in the school is 55 and the range for the school is 35 to 70.

(A, *go to* 7)
(B, *go to* 8)
(C, *go to* 6)

---

6 From the statements we can conclude that the student not only has a reading score lower than the mean for the school, but also has the lowest reading score in the school. Your answer, C, is correct; however, with a more sophisticated analysis of dispersion one could make further interpretations.

(*Go to* 9)

---

7 Although this example may be somewhat trivial, the logic of interpreting data is not. Answer A allows us only to know the child's reading score. Interpretation in this case requires comparing data or statistics in some way with a norm or standard. Choose another answer.

(*Go to* 5)

---

8 From statement B we can conclude that the student has a reading comprehension score lower than the school mean. However, information about the dispersion of scores in the school or other normative group would allow us to say more about the student compared to his classmates. Answer C should have been chosen.

(*Go to* 6)

---

9 One important characteristic of the range as a measure of dispersion is illustrated by the following two sets of data:

Set 1: 87, 89, 92, 95, 96, 97, 120
(Range = 87 to 120)
Set 2: 25, 86, 87, 88, 89, 90, 94
(Range = 25 to 94)

Think about the data values and the ranges. From these data one should conclude that the range is characterized by marked ________.

A. Stability
B. Instability

(A, *go to* 10)
(B, *go to* 11)

---

10 Look again at the data. Stability is not the correct answer. Note that if we removed the maximum score (i.e., 120) from set 1, and if we removed the minimum score (i.e., 25) from set 2, then the two ranges would become similar: range of set 1 = 87 to 97; range of set 2 = 86 to 94. Because a single extreme score can drastically alter the range, the range is said to be an unstable measure of dispersion. It is *sensitive to extremes.*

(*Go to* 12)

---

11 Because a single extreme score can drastically alter the range, the range is said to be an unstable measure of dispersion. Answer B, instability, is correct. Note how removing the maximum score from set 1 and the minimum score from set 2 makes the two ranges almost the same: 87 to 97 and 86 to 94.

(*Go to* 12)

---

12 Another way of looking at the instability of the range is to consider drawing several samples from a population. If any of the samples happen to have an extreme case, then the range for that sample will be much larger than in the other samples. The *sampling fluctuation* of the range

Continued.

is large. The larger the sample, the more likely you are to include an extreme case. The range is a statistic that is easy to understand and calculate, but it has many limitations.

An example of the use of the range can be found in McLaughlin et al. (1973). Analysis of urine samples for lead content were made over a 3½-year period. At the beginning of the study the range of lead content from various observations was 65-78 μg/liter of urine, whereas at the end of the study the range was 11-45 μg/liter of urine. The range at the end of the study appeared to be much larger than at the beginning, but the extremes were lower values at the end than at the beginning of the study.

(*Go to* 13)

---

13 If instead of looking at just the two extreme scores in a distribution, we look at the number of scores at or below any given point, then we are dealing with *percentiles.* A percentile is the percent of the total number of cases that falls *below* a given point on the scale. Thus at the 50th percentile, 50% of all cases are below this point; at the 82nd percentile, 82% of all the cases are below this point.

Percentiles are often used to compare an individual score with the distribution of scores in some population or norm group. Thus a pediatrician, finding that a child falls into the 50th percentile of height for his age group, knows that the child is exactly in the middle of the distribution for all children of that age.

In all distributions the 50th percentile is equivalent to the

A. Mean
B. Median
C. Mode

(A, *go to* 14)
(B, *go to* 15)
(C, *go to* 16)

---

14 The 50th percentile, by definition, is exactly in the center of any ordered distribution. This is also the definition of the median. The mean, an incorrect answer, would be at the 50th percentile only in symmetrical, bell-shaped distributions.

(*Go to* 17)

---

15 The 50th percentile is in the middle of the distribution, and therefore it is also the median of the distribution. The mean and mode would also be the 50th percentile only if the distribution were symmetrical and bell-shaped.

(*Go to* 17)

---

16 The mode is the value with the highest frequency of occurrence. Only when the distribution is symmetrical and bell-shaped will the mode be in the center. The correct answer is the median, which, like the 50th percentile, is defined as the value in the center of the distribution.

(*Go to* 17)

17 Percentiles are most frequently used when attempting to find the value of a variable that is at a given percentile position in your distribution. The procedures described below represent the logic behind the calculation of positional measures, including percentiles. Keep in mind that both the median and percentiles are positional measures on an ordered array of data.

With the distribution in Table 5-2, what was the age at which 50% of the population had died from coronary disorders? The following example will take you through the logic of this calculation.

Of the given cases, 50% must fall below the point at which the 50th percentile is located. In Table 5-2 how many cases fall below this point? ______________ cases.

(*Go to* 18)

TABLE 5-2. Deaths due to coronary disorders by age groups

		Cumulative frequency	
Age	*f*	*Counting up*	*Counting down*
85-89	2	146	2
80-84	4	144	6
75-79	10	140	16
70-74	15	130	31
65-69	30	115	61
60-64	30	85	91
55-59	25	55	116
50-54	10	30	126
45-49	6	20	132
40-44	4	14	136
35-39	3	10	139
30-34	3	7	142
25-29	2	4	144
20-24	2	2	146
	N = 146		

18 Table 5-2 has 146 cases; 50% of 146 cases is 73 cases ($\frac{146}{2} = 73$). Now by cumulatively summing all frequencies from the lowest interval up, you can determine that the 73rd case is contained in the interval 60-64. Thus you can say that the age interval in which the 50th percentile falls is 60-64 years. Usually one wants to be more precise about the age, and so a location within the 60-64 age interval must be determined.

Five ages are represented in this interval (60, 61, 62, 63, and 64), and there are 30 cases in this interval (f = 30), so theoretically you can divide the interval up into 30 small subintervals each with a value of 5/30 = 0.17, where 5 = the number of intervals between 59.5 and 64.5 (the real limits of the interval 60-64), and 30 = the frequency of cases in that interval.

We can now locate each of these 30 points at its appropriate subinterval. There are 55 cases below the interval containing the 50th percentile. The 56th case, which falls in the interval containing the 50th percentile, has the value of the lower limit of that interval plus 0.17 (5/30 = 0.17), or 59.5 + 0.17 = 59.67. Continuing this summing procedure indicates that the 57th case must have the value 59.67 + 0.17 = 59.84.

With this procedure, what is the exact age by which 50% of the deaths due to coronary disorder will have occurred? (Remember that you are looking for the value of the 73rd score.)

Age = ______________

When you are ready . . .

(*Go to* 19)

---

19 Your answer can be found by adding 0.17 to the lower true limit of that interval until you reach the 73rd score; or 59.5 + 0.17 = 59.67 for the 56th score, 59.67 + 0.17 = 59.84 for the 57th score, 59.84 + 0.17 = 60.01 for the 58th score, . . . + 62.39 + 0.17 = *62.56 for the 73rd score*. The 73rd score represents the middle position in the distribution of 146 persons who died of coronary disorders.

However, this is a rather complex and tedious way of solving for an exact score at a given percentile. This entire procedure can readily be placed in the following formula:

$$P_x = \begin{array}{l}\text{Lower real limit of interval}\\ \text{containing number you are}\\ \text{after}\end{array} + \left(\frac{\text{Number of added cases}^*}{\text{Frequency (number of cases in interval)}} \times \begin{array}{c}\text{Width}\\ \text{of}\\ \text{interval}\end{array} \right)$$

*where "number of added cases" = the position value in which you are interested (here, 73) minus the cumulative frequency up to, but not including, the interval in which the position value falls (here, 55).

Or for the score at the 50th percentile of the data presented in Table 5-2:

$$P_{50} = 59.5 + \left(\frac{73 - 55}{30} \times 5 \right) = 59.5 + \left(\frac{18}{30} \times 5 \right) = 62.5$$

This number means that half of this population had died of coronary disorders by age 62.5.

With this procedure, calculate the age representing the 25th percentile in Table 5-2. The 25th percentile is identical to the first quartile. If you need some help with the solution of this problem . . .

(*Go to* 20)

If you can solve the problem without additional input, do so and . . .

(*Go to* 22)

---

20 To solve this problem you have to find the following figures:

1. The interval containing the age that falls at the 25th percentile, or

$$\frac{N}{4} = \frac{146}{4} = 36.5\text{th score}$$

where 4 = 1/4 (25% of the sample).

2. The 36.5th score falls in the age interval 55-59, or using the cumulative frequency "counting up" column, there are 30 cases that fall at or below 54 and 55 cases that fall below 59; therefore the 36.5th case falls below 59 but above 55.

Now using the formula:

$$P_{25} = \text{Lower real limit} + \left(\frac{\text{Added cases}}{\text{Frequency}} \times \text{Interval width} \right)$$

$$P_{25} = 54.5 + \left(\frac{36.5 - 30}{25} \times 5 \right) = 55.8$$

Where:

54.5 is the lower real limit of the interval containing the 36.5th case (the case falling at the 25th percentile).
36.5 − 30 = 6.5 or the number of cases added to the cumulative frequency up to but not including the 25th percentile and needed to arrive at that score.
25 = the frequency of the interval containing the 25th percentile.
5 = the width of the interval.

Therefore by the age of 55.8 years, 25% of the sample had died.

(*Go to* 21)

---

21 Solve for the age at which 75% of the sample found in Table 5-2 had died from coronary disorders.

When you have completed your calculations . . . (*Go to* 23)

---

22 The score at the 25th percentile is determined by use of the following formulas:

$$\frac{N}{4} = \frac{146}{4} = 36.5\text{th score, which falls in the interval } 55 - 59$$

$$P_{25} = 54.5 + \left(\frac{36.5 - 30}{25} \times 5\right) = 55.8$$

Therefore the 25th percentile falls at age 55.8. By the age of 55.8 years, 25% of the sample had died.

If the solution does not agree with yours or if you do not understand the solution . . .

(*Go to* 20)

Otherwise . . . (*Go to* 21)

---

23 1. 75% of N $= \frac{146}{4} \times 3 = 109.5$th score.

2. The 109.5th case falls in the 65-69 interval.
3. Applying the formula:

$$P_{75} = 64.5 + \left(\frac{109.5 - 85}{30} \times 5\right) = 64.5 + 4.1 = 68.6$$

In other words, by the age of 68.6 years, 75% of the sample had died from coronary disorders.

(*Go to* 24)

---

24 The method for finding the value of any percentile just described is the method used when you have grouped data or when you want to find a relatively precise value for a percentile, decile, median, or other positional measure. In practical statistics you may want to simply indicate the category or midpoint of a category when communicating with a colleague. Thus in Table 5-2 you could refer to the median as the midpoint of the interval in which it falls (62) rather than the more precise 62.56 that we calculated. In Table 5-3 the first quartile of the vocabulary scores is 16; or, 25% of the vocabulary scores fall below 16. It is simpler to refer to the first quartile in this manner than to calculate the fractional score within the category, since fractional scores cannot actually be obtained and may lead to confusion in interpretation. Furthermore, much of the data used in the conduct of research are in the ungrouped, or raw, form found in Table 5-3.

If you understand how the score 16 was arrived at . . . (*Go to* 26)

If you are unsure as to how 16 was obtained . . . (*Go to* 25)

TABLE 5-3. Distribution of sixth grade vocabulary scores in Morton School

Raw score	*f*	*Cumulative f*	*Raw score*	*f*	*Cumulative f*
10	1	1	18	7	31
11	0	1	19	3	34
12	2	3	20	6	40
13	1	4	21	4	44
14	4	8	22	2	46
15	3	11	23	0	46
16	5	16	24	1	47
17	8	24	25	1	48
N = 48					

25 There are 48 pupils' scores represented in Table 5-3, and the first quartile (Q_1 or P_{25}) is represented by $\frac{N}{4}$, or $\frac{48}{4}$ = the 12th score. There are 11 scores up to the raw score of 15, so the 12th score must fall at the raw score value of 16.

(*Go to* 26)

26 Percentiles are frequently encountered in literature in applied fields. Percentiles divide the total range into 100 parts, each of which has the same number of cases. Less commonly you may find references to deciles or quartiles. Deciles divide the total range into ten parts, each of which contain ten percentiles. Quartiles are the percentiles that divide the distribution into quarters—the 25th, 50th, and 75th percentiles are the 1st, 2nd, and 3rd quartiles. The 1st, 2nd, and 3rd quartiles are commonly symbolized as Q_1, Q_2, and Q_3.

Frequently we find percentiles, or variations of percentiles such as quartiles, used to divide data into various categories for use in analyzing certain variables. An example of this is seen in the work of Kasl et al. (1970), where achievement was measured by the student's grade-point average. To analyze the relationship between achievement and various physiologic measures, grade-point average was divided into four categories based on quartiles:

$$Q_1 \leq 1.89;\ Q_2 = 1.90 \text{ to } 2.29;\ Q_3 = 2.30 \text{ to } 2.79;\ \text{and } Q_4 \geq 2.80$$

The authors found that as grade-point average increased (from Q_1 to Q_4), the level of uric acid also increased.

(*Go to* 27)

27 Another measure of dispersion is the *semi-interquartile range,* which is simply a variation of the range. The semi-interquartile range, Q, is defined as half the range between the 1st and 3rd quartiles (i.e., half the range of the middle 50% of the cases).

Stated algebraically:

$$Q = \frac{Q_3 - Q_1}{2}$$

Determine Q for the data in Table 5-2 (you have already calculated Q_1 and Q_3 in a previous exercise).

$$Q_1 = P_{25} = 55.8 \qquad Q_3 = P_{75} = 68.6$$

When you have completed the calculation of Q . . .

(*Go to* 28)

28 The solution for Q is $\frac{68.6 - 55.8}{2} = 6.4.$

The semi-interquartile range, by eliminating extreme scores, varies less from sample to sample than does the range, and therefore it is a more stable measure of dispersion. However, it does not take into account all of the information (i.e., the scores less than 25% and more than 75%). Although it is a crude measure of the extent of variability, it is most frequently used when comparing one sample with another, or when comparing two groups receiving different types of treatment procedures, as in the following example.

Two groups of new residents who moved into town X 3 years ago are being studied for their voter turnout over the past 3 years. One group has been mobile, moving at least four times in the past 10 years. The other group has been fixed, moving only once in the last 10 years. For exploratory purposes we want to determine which group shows the most consistent voter participation.

Mobile group: $Q_1 = 6$ elections
$Q_3 = 14$ elections
Fixed group: $Q_1 = 7$ elections
$Q_3 = 10$ elections

If other factors are constant, which group shows a more consistent voter turnout in number of elections in which they participated?

Perform your calculations and then . . .

(*Go to* 29)

29

$$Q_{\text{mobile}} = \frac{14 - 6}{2} = 4$$

$$Q_{\text{fixed}} = \frac{10 - 7}{2} = 1.5$$

The difference between the two semi-interquartile ranges indicates that the more fixed population shows more consistency—less than half the variation in voter turnout compared to the mobile population. There is more dispersion in the mobile population.

It must be noted that the semi-interquartile range does not indicate which group votes most frequently. For this information some measure of central tendency is needed. All we know from the Q performed on these data is that the less mobile group is more consistent in their voting behavior than is the mobile group.

In the "normal" or bell-shaped curve the mean and the median fall at the same point on the distribution. In a normal curve, if the value of Q is added to and subtracted from the median, that part of the curve between the median +1Q and the median −1Q includes __________% of the cases.

(Go to 30)

30 That part of the normal curve between the median +1Q and the median −1Q includes 50% of the cases. Fig. 5-2 demonstrates the relationship between Q and the median.

(Go to 31)

FIG. 5-2

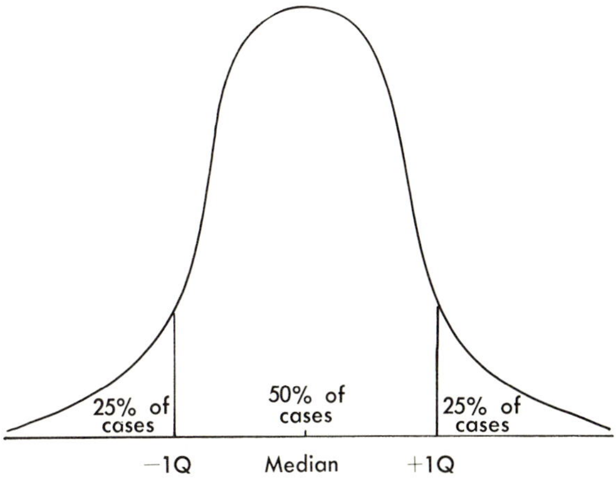

31 In this manner, Q identifies the middle half of a distribution as well as the upper 25% and the lower 25% through a process of subtraction.

You have seen what happens to the range when an extreme score is included in the distribution. An extreme score (does/does not) have the same effect on the semi-interquartile range.

A. Does (A, *go to* 32)
B. Does not (B, *go to* 33)

32 Your answer, A, an extreme score has the same effect on Q as it has on the range, is incorrect for several reasons. The fact that Q is a range divided by 2 means that any effect on the range is immediately halved. Also, the interquartile range, $Q_3 - Q_1$, is highly dependent on frequency, so that intervals with low frequency, as is usually the case with extreme scores, have limited effect on the interquartile range. Thus Q is much more stable than the range as a measure of dispersion. (*Go to* 34)

33 An extreme score does not have the same effect on Q as it has on the range; your answer is correct. Q is a range divided by 2; therefore the effect that extreme scores have on a range is immediately halved. Also Q_3 and Q_2 are highly dependent on frequency; extreme scores, which are usually of a low frequency, have a limited effect on the two quartiles. (*Go to* 34)

34 Given your knowledge of the range and Q, it is now possible to begin a discussion of the *standard deviation* and the normal curve. The standard deviation is the most widely encountered measure of dispersion in most research using statistical procedures.

To understand the concept on which the standard deviation is based, it is first important that more information about the normal curve be presented. Since the normal curve is a symmetrical distribution, there must be 50% of the distribution on one side of the curve and 50% of the distribution on the other; that is, the mean (median and mode also) divides the normal distribution in half.

Through the use of calculus, it is possible to determine exact areas between any two points under the normal curve. The usefulness of determining a standard set of areas will become apparent later when probability and significance testing are introduced. In statistics, the standard deviation defines a distance away from the mean or center of the normal curve such that in all normal curves, regardless of mean or dispersion, the area or proportion of cases is constant (or standard). Thus in all normal curves, the proportion of cases (area under the curve) between the mean and 1 standard deviation (or 2 or 3 standard deviations) will always be the same. The notion of constant areas under the normal curve is an important concept and should be internalized by the learner before going on. (*Go to* 35)

---

35 The standard deviation is a kind of average of all deviations about the mean. A more exact explanation will be given below. The symbols used for standard deviation are either s or σ. The symbol s is used when the reference is to the standard deviation of a sample, whereas σ is used when the reference is to the standard deviation of a population.

Look at Fig. 5-3. All normal curves have those constant proportions between each standard deviation; between the mean (or median or mode) and 1 standard deviation on either side of the curve, you will find 34.13% of the total curve. In other words, 34.13% of all scores represented by this curve will always lie between the mean and 1 standard deviation above or below the mean.

The area between the mean and $+1s = 34.13\%$, and the area between the mean and $-1s = 34.13\%$. Therefore the area $\pm 1s = 68.26\%$ or slightly over two thirds of the total normal curve area.

What proportion of the normal curve is included in the area between $\pm 2s$?

A. 27.18% (A, *go to* 38)
B. 47.72% (B, *go to* 37)
C. 95.44% (C, *go to* 36)

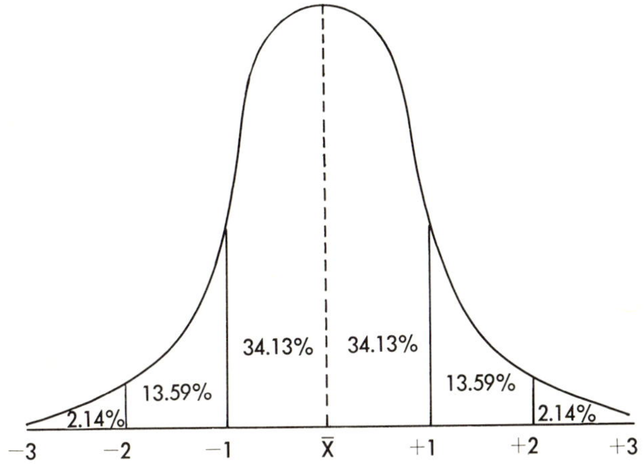

FIG. 5-3. Normal curve with ±1, ±2, and ±3 standard deviations around the mean.

36 The area of the curve between ±2s is in fact 95.44% of the curve, since it includes all of the area from the mean to +2s plus the area from the mean to −2s. (See Fig. 5-3.) You should try to remember the area of a normal distribution within ±1s and ±2s because it provides a convenient way of figuring where the bulk of a sample falls on some variable (if you can assume a reasonably normal distribution).

(*Go to* 39)

37 You have calculated the area under the curve between the mean and 2s as 47.72%. Your answer is not correct.

(*Go to* 36)

38 The area between +1s and +2s is 13.59%. Adding another 13.59% for the area between −1s and −2s totals 27.18%. But this area does not include the area between the mean and +1s, which is part of the area originally asked for.

(*Go to* 36)

39 If you add all of the areas under the curve presented in Fig. 5-3, you will discover that they total 99.72% instead of 100%. The normal curve is a form of a theoretical frequency distribution, and as such it extends out to infinity in both directions (note that neither end of the curve touches the horizontal line). However, we speak of the total area under the curve as being equal to 100%.

(*Go to* 40)

40 The formula for the standard deviation helps explain this statistic.

$$\sigma = \sqrt{\frac{\Sigma(X - \overline{X})^2}{N}}$$

$(X - \overline{X})$ is the deviation of a score from the mean of the group. The mean is the reference point for our measure of dispersion. We square the deviation, which magnifies the effect of extreme values and also avoids problems with negative deviations. Then we sum (Σ) the squared deviations, which gives us the sum of squares. Dividing by N gives us the mean square. Because of the earlier squaring process, we take the square root of the mean square. Because of the squaring and square root procedures, we do not have the mean deviation, but we have something close to it. What we have is the square root of the mean of the squared deviations from the mean, which is called the *standard deviation.*

Frequently the formula for the standard deviation appears as

$$\sigma = \sqrt{\frac{\Sigma x^2}{N}}$$

where x is a shorthand form of $(X - \overline{X})$. The replaceability of these symbols should not confuse you.

(*Go to* 41)

---

41 *Note 1*

In many texts you will find the formula for s as:

$$s = \sqrt{\frac{\Sigma(X - \overline{X})^2}{N - 1}}$$

The difference between this formula and the one in the previous frame is that the denominator is N − 1 instead of N. The effect of dividing by N − 1 instead of N is to increase slightly the value of s. This increase is a conservative act (it leads to fewer errors in conclusions) and provides a less biased estimate of σ when we are interested in estimating the population standard deviation (σ) from the sample standard deviation (s). When N is large, the effect of using N − 1 instead of N becomes trivial. For this text and for most practical problems you should feel free to use N in the denominator (i.e., the formula in frame 40).

(*Go to* 42)

---

42 *Note 2*

Measures of dispersion are also called measures of variability. You also will encounter the term *variance*. Variance, in statistics, means something specific: it is the mean of the squared deviations about the mean.

$$\text{Variance} = \frac{\Sigma(X - \overline{X})^2}{N}$$

As you may notice:

$$s = \sqrt{\text{Variance}}$$

That is, the standard deviation is the square root of the variance, or, alternatively, variance is the standard deviation squared.

(*Go to* 43)

---

43 Table 5-4 presents a series of data and some statistical manipulations of these data.

Using the formula

$$s = \sqrt{\frac{\Sigma(X - \bar{X})^2}{N}}$$

calculate the standard deviation. A table of square root values is presented in Appendix A, Table 1.

When you are done . . . (*Go to* 44)

TABLE 5-4. Deviation from mean

Scores X	*Deviation from mean* $(X - \bar{X})$	*Deviation squared* $(X - \bar{X})^2$
8	(8 − 5) = 3	3 × 3 = 9
7	(7 − 5) = 2	2 × 2 = 4
6	1	1
5	0	0
4	−1	1
3	−2	4
2	−3	9
$\Sigma X = 35$ $\bar{X} = 5$ $N = 7$	$\Sigma(X - \bar{X}) = 0$	$\Sigma(X - \bar{X})^2 = 28$

44 Your calculations should be as follows:

$$s = \sqrt{\frac{\Sigma(X - \bar{X})^2}{N}} = \sqrt{\frac{28}{7}} = \sqrt{4} = 2$$

The standard deviation is 2.

Standard deviations are expressed in terms of the units of the variable X. There is no true verbal interpretation of the figure that you have arrived at through your calculations. The size of the standard deviation must be interpreted in terms of the range and mean of the scores and frequently in comparing one distribution with another distribution. For the data in Table 5-4 with a range of 2-8 and a mean of 5, the standard deviation is fairly large. Had the standard deviation been 1, it would have represented less deviation about the mean.

You will recall from the discussion of the properties of the mean from the last section that the deviation of all of the scores about the mean will always be zero. Since the sum of the column $(X - \bar{X})$ will always be zero, (Table 5-4), this is a useful way of checking your work.

Table 5-5 represents a set of data and a "dummy table" for the calculation of the standard deviation.

Calculate s for these data and then . . . (*Go to* 45)

TABLE 5-5. Number of trials required to learn maze by "maze-dull" rats

Raw scores: 18, 16, 14, 12, 10, 8, 6

X	$(X - \bar{X})$	$(X - \bar{X})^2$
$\Sigma X =$ $\bar{X} =$	$\Sigma(X - \bar{X}) =$ $N =$	$\Sigma(X - \bar{X})^2 =$

45 The correct calculation for Table 5-5 is as follows:

X	$(X - \bar{X})$	$(X - \bar{X})^2$
18	6	36
16	4	16
14	2	4
12	0	0
10	−2	4
8	−4	16
6	−6	36
$\Sigma X = 84$ $\bar{X} = 12$	$\Sigma(X - \bar{X}) = 0$ $N = 7$	$\Sigma(X - \bar{X})^2 = 112$

$$s = \sqrt{\frac{\Sigma(X - \bar{X})^2}{N}}$$

$$s = \sqrt{\frac{112}{7}} = \sqrt{16}$$

$s = 4$ trials

Obviously, the task of calculating a standard deviation without the aid of a computer or desk calculator can be tedious. Even with desk calculators, using the definitional formula presented can be time-consuming. Therefore through simple mathematical derivation, the following formula (called a computational formula) can be derived for simplifying the task of calculating the standard deviation.

$$s = \sqrt{\frac{\Sigma X^2}{N} - \bar{X}^2} = \frac{1}{N}\sqrt{N\Sigma X^2 - (\Sigma X)^2}$$

Using the data in Table 5-5, recalculate the standard deviation using one of the computational formulas.

To check your work . . .

(*Go to 46*)

46

X	X^2
18	324
16	256
14	196
12	144
10	100
8	64
6	36

$\Sigma X = 84 \quad \Sigma X^2 = 1120 \quad N = 7$

$\overline{X} = \frac{\Sigma X}{N} = \frac{84}{7} = 12$

$$s = \sqrt{\frac{\Sigma X^2}{N} - \overline{X}^2}$$

$$s = \sqrt{\frac{1120}{7} - (12)^2}$$

$$s = \sqrt{160 - 144}$$

$$s = \sqrt{16} = 4 \text{ trials}$$

or

$$s = \frac{1}{N}\sqrt{N\Sigma X^2 - (\Sigma X)^2}$$

$$s = \frac{1}{7}\sqrt{7(1120) - 84^2}$$

$$s = \frac{1}{7}\sqrt{7840 - 7056}$$

$$s = \frac{1}{7}\sqrt{784}$$

$$s = \frac{1}{7}(28) = 4 \text{ trials}$$

The choice of which of the three formulas to use should be based on what you find to be the easiest, since:

$$s = \sqrt{\frac{\Sigma(X - \overline{X})^2}{N}} = \sqrt{\frac{\Sigma X^2}{N} - \overline{X}^2} = \frac{1}{N}\sqrt{N\Sigma X^2 - (\Sigma X)^2}$$

The interpretation of $s = 4$ is based on its relationship to the mean and the range. Where $\overline{X} = 12$ and the range is from 6 to 18 trials, a deviation of 4 trials is not large.

For more practice in computing a standard deviation . . . *(Go to 47)*

Otherwise . . . *(Go to 48)*

47 Use a computational formula to calculate the standard deviation for the following data: 14, 10, 8, 6, 6, 3, 1.

Finish your calculations, then check your work below.

X	X^2
14	196
10	100
8	64
6	36
6	36
3	9
1	1

$\Sigma X = 48 \quad \Sigma X^2 = 442$

$N = 7$

$$s = \sqrt{\frac{\Sigma X^2}{N} - \overline{X}^2}$$

$$s = \sqrt{\frac{442}{7} - \left(\frac{48}{7}\right)^2}$$

$$s = \sqrt{63.1 - (6.9)^2}$$

$$s = \sqrt{63.1 - 47.6}$$

$$s = \sqrt{15.5} = 3.9$$

(Go to 48)

48 A few examples of the use of the standard deviation may be helpful here. Toomey (1972) questioned subjects who had gone through two forms of alcohol detoxification programs run by a mental hospital. He found that the experimental subjects (those going through a group experience while hospitalized) reported a mean wellness ("do you feel much worse"; "do you feel much better") of 6.882 with a standard deviation of $s = 0.332$. Those subjects who were in a control group (no group experience) reported an $\overline{X}$ wellness of 5.667 and $s = 2.422$. The range of scores was 1 to 7. Thus those going through the group experience not only felt better, but also deviated around this higher mean much less than those not undergoing the group experience (0.322 vs. 2.442). If the scores on the "wellness" questionnaire are normally distributed, then of those subjects going through the experimental situation, 68.26% should have scores between 6.55 and 7.21, whereas those not in the experimental group should have approximately two thirds of the scores between 3.245 and 8.089. Recall that the 68.26% figure comes from the fact that ± 1 standard deviation on the normal curve always includes that exact percentage of all cases.

Booth (1972) found, while studying friendship, that although widowed women had more friends that did widowed men, the variability as measured by the standard deviation was greater for women than men ($s_{men} = 2.4$; $s_{women} = 3.2$). Thus women varied more in their responses in Booth's research than did men.

(Go to 49)

49 To reiterate, the standard deviation is a *statistic* that *describes* the extent of the dispersion of the scores about the mean. The interpretation of this statistic is based on the amount of area that a certain number of scores occupy under the normal curve. More detailed use of the standard deviation will be made later; however, the standard deviations you calculated are critical statistics in the process of tests of statistical significance, as well as being effective for describing data.

When interpreting data, it is important to relate measures of dispersion to other statistics in a problem situation. For instance, if you are comparing the effects of two treatments (e.g., instructional programs, leadership styles, or drugs), you need to consider the amount of effect (change or gain) as well as the dispersion of the effect and where the two groups were to begin with. A drug that produces a larger mean effect but with great variability may be less desirable than a drug that produces a smaller mean effect but with great consistency. One's decision as to which drug to use in a given circumstance would depend not only on the statistical results, but also partly on factors such as risk, cost, and possibilities for changing the treatment later.

(End of chapter 5)

Examination for chapters 1 to 5

To test the adequacy of your understanding of the content material to this point in the book it is recommended that you take the following examination. The examination was devised by turning most of the learning objectives at the beginning of each chapter into questions. If you met the objectives outlined, then you should have no trouble with the examination.

If the question requires the calculation of a statistic, look up the formula in the text or in Appendix C. Otherwise, do not use your text. Appendix B contains the answers to the questions. If you answered a question incorrectly, the answer has a reference to a specific chapter and learning objective. You should go back to that chapter and the appropriate frames to restudy that particular material.

GOOD LUCK

1. Classify the following statements as indicative of descriptive or inductive statistical statements:

   A. A psychologist uses students from a large lecture class to provide data for an experiment on color vision and finds that 17% of the men are color-blind.

      A. Descriptive _____
      B. Inductive _____

   B. A survey of 1924 households in the state of Missouri indicates that 76% of all Missouri households have physicians who are considered their "family doctors."

      A. Descriptive _____
      B. Inductive _____

2. Match the following concepts with the definitions:

Concepts	*Definitions*
A. Sample ____	1. Manipulated measurements of population or sample
B. Population ____	2. A subset of a larger whole
C. Data ____	3. A measurement or unit of observation that takes on different values
D. Statistic ____	4. All units of observation in a specified area having at least one characteristic in common
E. Variable ____	5. Measurements of some population or sample

3. Name the various levels of measurement in ascending order of complexity.
4. For each of the following variables, state the highest level of measurement associated with it.
   A. Social class classified as upper, middle, lower
   B. Income classified as dollar amounts
   C. Age classified as chronological years
   D. Health classified as good, fair, poor
   E. Sex classified as male, female
   F. Race classified as black, white, other
5. For the data in the table below:
   A. Relative to other symptoms, headache symptoms were reported with the greatest proportion in which age group?
   B. Among respondents 40 to 59 years of age, 3 of 7 reported which symptom?
   C. Is the ratio of headache to exhaustion in the 18- to 39-year-old group greater or less than that in the 40- to 59-year-old group?

Distribution by age of psychosomatic illnesses reported by study sample

		Age	
Symptom	*18-39*	*40-59*	*60 and over*
Headache	16	30	10
Exhaustion	10	16	11
Nervousness	6	24	14

6. Prior to an experiment involving the cooperation of strangers, a measurement of hostility was taken on each subject. Subjects were subsequently randomly divided into two groups: group 1 received instructions for the experiment from the experimenter that were

confusing and contradictory and group 2 received instructions that were clear and unambiguous. From the data presented below, and using the formula for rate of change found in chapter 2, determine which group's hostility index changed the most and interpret the differences in rates of change.

	Hostility index	
	Before	*After*
Group 1 (confused instructions)	8	13
Group 2 (clear instructions)	10	8

7. A university is interested in the ages of persons entering the junior year at that institution. What follows is an array of data from a sample of students at the university who are enrolled as juniors.

20	19	19	20	20	25
21	27	20	21	23	21
18	20	18	22	21	20
20	18	19	19	20	20
20	19	21	20	29	26

From these data:

A. Determine the range.

B. Construct a frequency table with an interval equal to size 2 and starting at the lowest number in the array.

C. Compute percentages for each interval.

D. Compute the cumulative frequency and cumulative percentages (up).

E. Is the resulting curve drawn from these data left- (negatively) or right- (positively) skewed?

8. Draw three separate curves, one that is platykurtic, one that is mesokurtic, and a third that is leptokurtic.

9. The standard deviation, range, and semi-interquartile range are called measures of ________________.
The mean, median, and mode are called measures of __________.

10. Match the following concepts with the definitions.

Concepts	*Definitions*
A. Median	1. Sum of all values divided by the number of values
B. Mean	2. Half the range of the middle half of the cases
C. Semi-interquartile range	3. Value having greatest frequency
D. Mode	4. Center value
E. Standard deviation	5. Square root of the mean of the squared deviations from the mean

11. From the raw data presented in problem 7:
    A. Calculate the mode.
    B. Calculate the median.
    C. Calculate the mean.
    D. Using only the consideration of level of measurement, which of the above measures of central tendency would be most appropriate for these data?
    E. What is the minimum level of measurement required for each of the three measures of central tendency you just calculated?
    F. Discuss the effect of the skewness of the distribution on the three measures of central tendency.
    G. Which of the three measures would be "best" for this distribution?

12. From the frequency distribution calculated in problem 7 above:
    A. Locate and interpret the 75th percentile.
    B. Locate and interpret the 25th percentile.
    C. Calculate the semi-interquartile range.
    D. Interpret the difference between the range and the semi-interquartile range.

13. Given a normal distribution of scores on a standardized examination:
    A. What proportion of cases will fall between $\pm 1s$?
    B. What proportion of cases will fall between $\pm 2s$?
    C. If the exam takers in group I have a mean of 75 and approximately two thirds of the scores are from 70 to 80, whereas group 2 has an identical mean but two thirds of the scores are from 72 to 78, which group has the larger standard deviation, and what are the two values?

14. Given the raw data in problem 7, calculate the standard deviation.

15. A congressional state representative was interested in ascertaining the attitudes of his constituents regarding the Equal Rights Amendment for women. He contracted with a social scientist to conduct a survey of 250 men and 250 women randomly selected from his district. On a scale from 0 to 10, where 0 = opposed to ERA and 10 = in favor of ERA, the following data were obtained.

Men	*Women*
$\overline{X}_M = 5.9$	$\overline{X}_F = 6.1$
$\text{Median}_M = 5.9$	$\text{Median}_F = 6.1$
$s_M = 3.4$	$s_F = 2.1$

Interpret these findings.

16. Match the following symbols with the list of terms.

Symbols	*Terms*
A. Q	1. Semi-interquartile range
B. σ	2. Percent
C. %	3. Frequency
D. Σ	4. Standard deviation (population)
E. s	5. Mean of sample
F. $\overline{X}$	6. Size of sample or population
G. N	7. Standard deviation (sample)
H. f	8. Summation

chapter
6

Basic probability

At the conclusion of this chapter, you should be able to:

1. Label the ratio: $\frac{\text{Number of events of interest}}{\text{Total possible events}}$ as a probability statement for equally likely events.
2. Define probabilities (statements of likelihood about a population expressed as values from 0 to 1).
3. Use the addition theorem of probability to the extent of calculating the probability of rolling a 1 or a 6 on one roll of a die.
4. Use the multiplication theorem of probability to the extent of calculating the probability of rolling a 1 and a 6, in that order, on two rolls of a die.
5. Use the addition and multiplication theorems combined to the extent of calculating the probability of rolling a 1 and a 6, in any order, on two rolls of a die.
6. Recognize the problems of independence and sampling with or without replacement to the extent of being able to calculate the difference in probabilities of drawing two blondes in two draws out of a population of four blondes and six brunettes (a) with replacement and (b) without replacement.
7. Correctly apply the label "independent events" to a set of examples.

---

1 Probability theory is the basis for decision-making or for making statistical inferences. Instead of saying: if A is true, then B must follow, the statistical inference is: if A is true, then within certain limits of probability, B will follow.

Although the application of probability concepts to inductive sta-

Continued.

tistics may not be readily apparent, the student of statistics should be assured of the fact that a firm grasp of basic probability theory will be necessary to understand the nature of tests of statistical significance and the application of probabilistic hypotheses.

How do we know that instructional method A is better than instructional method B? How do we know that a set of symptoms is an indicator of a particular disease? We expect the answers we give because in the past we have observed data. If we are rigorous in our method of data collection and analysis, we may get a different answer than we will if we allow our mind to focus in on a particular example. But in either case, we make a *probability* statement that "based on theory or experience, we expect the following." Indeed, nearly all our behavior is based on expectations generated from experience. We expect the floor to be there when we get out of bed, and we expect pain to follow a kick in the shins. These expectations are probability statements based on inferences from prior data.

(*Go to* 2)

---

2 A simple example may serve to illustrate the probabilistic nature of hypothesis testing. Suppose an obstetrics nurse claims she can tell the sex of an 8-month human fetus by listening to the heart beat, and sure enough she is right on the next case. Since she could do that just by chance, we test her again and she is again correct. Now the question arises of what the chances are of her being correct by chance in two trials. You may be able to calculate that the chances are 1 in 4 of being correct on two trials. But suppose she is correct on three trials, or four, or six out of eight. We would want to know the probability of her being correct by chance before we decided on her claimed ability to determine the sex of unborn infants.

Generally, we know that a set of symptoms is an indicator of a particular disease because:

A. A given disease has symptoms that are not reflected in any other disease.

B. With today's instruments, measurement of abnormal body functions is easily detected.

C. Other people who were successfully diagnosed as having the same disease also had the same set of symptoms.

(A, *go to* 3)
(B, *go to* 4)
(C, *go to* 5)

---

3 Some symptoms may indicate other diseases (e.g., a high temperature presents itself in a large array of specific diseases). However, if the same set of symptoms appeared with other people who were successfully diagnosed with the same disease, there could be greater faith in using that set of symptoms as an indicator of that disease.

(*Go to* 6)

4 Instruments will measure abnormalities of body functions, but whether the abnormalities indicate the disease is a question still unanswered. If it were known that these are symptoms that other people have had who were successfully diagnosed as having the same disease, there would be a basis for using these symptoms as an indication of the disease.

(*Go to* 6)

5 Other people who were successfully diagnosed with the same disease also had the same set of symptoms. Correct. It is because of this fact that we know that a set of symptoms is indicative of a particular disease.

(*Go to* 6)

6 Because numerous individuals with the same disease experienced the same set of symptoms, it is probable that future cases with the same symptoms will have the same disease. Such reasoning represents one characteristic of a probability statement. Probability refers to a population—to a large number of experiences, events, or the outcome of things that will happen in the long run. The terms "likelihood" and "chance" are also often used.

For example, in the long run, the probability for flipping heads of a coin is:

A. Equal to the probability of flipping tails
B. Less than the probability of flipping tails

(A, *go to* 8)
(B, *go to* 7)

7 Assuming a "fair" coin, a head is just as likely to land up as a tail. Your answer is incorrect. In the long run, the probability for flipping heads of a coin must be equal to the probability of flipping tails, since both are equally likely outcomes.

(*Go to* 8)

---

8 The probability for heads would be the same as for tails. In the long run, for example 1000 separate flips of the same unbiased coin, the chances are good that there would be close to 500 heads and 500 tails. From this, it can be surmised that probability can be calculated when:

A. The total number of possible outcomes are known (two kinds—heads or tails)
B. The number of times the desired outcome can appear is known (heads—once)
C. Both of the above

(A, *go to* 9)
(B, *go to* 10)
(C, *go to* 11)

---

9 Actually, both the total number of outcomes possible and the number of times the desired outcome can appear need to be known. Thus, in the case of the coin, the possible outcomes are two. There can possibly be heads or tails when a coin is flipped, but only one of these at a time. The number of desired outcomes is one for the coin. Only heads are desired. Therefore the probability for flipping heads on any one flip of a coin is 1/2 (or .5) because there is one chance for what is desired (heads) out of two possible outcomes. In dice the total number of possible outcomes of rolling a die is six. It could land with any one of six possible faces facing up.

(*Go to* 12)

---

10 It is true that you must know the number of times the desired outcome can appear. In the case of the coin, the number is one, because with one flip of the coin you desire only one kind of outcome or success—heads. However, to calculate the probability of getting a head on one flip, you also need to know how many possible ways the coin could land. Since a coin has two sides, this number must be 2. The probability, then, can be determined as being 1/2 (or .5). There is one chance for what is desired out of two possible outcomes. If a die were rolled, the total possible outcomes (values) for any one roll of the die would be six.

(*Go to* 12)

---

11 Both items of information are needed to calculate the probability for flipping a head on any one flip of the coin. Only one kind of outcome (heads) is desired, and there are only two possible outcomes. Therefore the probability for getting heads is 1/2 (or .5). There is one chance for the desired outcome out of two possible outcomes.

If one were rolling dice, the total possible different outcomes (values) for any one roll of a die would be six.

(*Go to* 12)

---

12 If we desire to get a 2 when we roll one die, what are our chances for success?

A. 1/2
B. 1/6
C. 2/6

(A, *go to* 13)
(B, *go to* 14)
(C, *go to* 15)

---

13 A die has six possible ways it could land. You are interested in only one side, the side with the 2. Therefore, when the die is rolled, there is one chance in six that a desired outcome will turn up. Mathematically, the formula for finding the probability for equally likely events is the ratio:

$$\frac{\text{Number of events or outcomes of interest}}{\text{Total possible events or outcomes}} = P$$

$$1/6 = P$$

(*Go to* 16)

---

14 There is, as you said, one chance in six that a 2 will turn up—if the die is an honest one.

The mathematical formula for finding the probability for equally likely events is the ratio:

$$\frac{\text{Number of events or outcomes of interest}}{\text{Total possible events or outcomes}} = P$$

$$1/6 = P$$

(*Go to* 16)

---

15 Although there are six possible ways the die could land, only one way would satisfy our interest, a 2. Therefore the probability would be only one chance in six that a 2 would be rolled—if the die is an honest one. The mathematical formula for finding the probability for equally likely events is the ratio:

$$\frac{\text{Number of events or outcomes of interest}}{\text{Total possible events or outcomes}} = P$$

$$1/6 = P$$

(*Go to* 16)

---

16 What would be the probability of randomly drawing a 6 from a box of nonrepeated numbers that are written on separate pieces of paper?

A. 6/30
B. One chance in six
C. More information is needed

(A, *go to* 17)
(B, *go to* 17)
(C, *go to* 18)

---

17 The probability cannot be calculated unless the number of pieces of paper in the box is known. In the example just cited, the total number of possible outcomes is unknown. More information is needed.

(*Go to* 19)

---

18 More information is needed. It is not known how many pieces of paper are in the box. If one knew the total number of possible outcomes, the probability could be calculated.

(*Go to* 19)

---

19 The formula for finding the probability of any set of equally likely events is:

A. $\frac{\text{Number of outcomes or events of interest}}{\text{Total possible outcomes or events}} = \text{P}$

B. $\frac{\text{Total possible outcomes or events}}{\text{Number of outcomes or events of interest}} = \text{P}$

C. $\frac{\text{Number of times trials are repeated}}{\text{Number of desired outcomes or events}} = \text{P}$

(A, *go to* 20)
(B, *go to* 21)
(C, *go to* 22)

---

20 You chose the correct formula:

$$\text{P} = \frac{\text{Number of outcomes or events of interest}}{\text{Total possible outcomes or events}}$$

(*Go to* 23)

---

21 You chose the wrong formula. The numerator and denominator should be reversed. The formula for finding the probability of any set of equally likely events is:

$$\text{P} = \frac{\text{Number of outcomes or events of interest}}{\text{Total possible outcomes or events}}$$

(*Go to* 23)

---

22 You chose an incorrect formula. The number of times the coin is flipped or die is thrown will not affect the probability ratio for obtaining a desired outcome or event. The correct formula is:

$$P = \frac{\text{Number of outcomes or events of interest}}{\text{Total possible outcomes or events}}$$

(*Go to* 23)

23 Probabilities are statements of likelihood expressed as values from 0 to 1. One can readily see the mathematical reason for this range of values from examining the formula.

$$P = \frac{\text{Number of outcomes or events of interest}}{\text{Total possible outcomes or events}}$$

The number of outcomes of interest can never exceed the total possible outcomes, and the resulting probability will always be a fraction.

What would be the likelihood of throwing a 2 *or* 5 in one throw of a die?

A. 1/6 (A, *go to* 24)
B. 1/3 (B, *go to* 26)
C. 1/12 (C, *go to* 25)

24 You have to remember there are two outcomes of interest—getting either a 2 or 5. The total number of possible outcomes is still six.

(*Go to* 25)

25 The probability for throwing a 2 would be 1/6, and the probability for throwing a 5 would be 1/6. The probability for throwing a 2 *or* a 5 in any one throw would be 2/6 or 1/3. If events A and B (throwing a 2 or 5) are *mutually exclusive*, that is, if one occurs, then the other cannot possibly occur at the same time, then the probability of getting either A *or* B is equal to the probability of A *plus* the probability of B. What would be the probability of drawing an ace *or* a king from a deck of 52 cards? (*Go to* 27)

26 You are correct that the probability for a 2 or 5 on one throw would be 1/6 + 1/6 or 2/6 = 1/3. Thus if events A and B are *mutually exclusive*, that is, if one occurs, then the other cannot possibly occur at the same time, the probability of getting either A *or* B is equal to the probability of A *plus* the probability of B. What would be the probability for drawing an ace *or* a king from a deck of 52 cards? (*Go to* 27)

27 The answer is P = 8/52 = 2/13.

There are four aces in the deck of 52 possible cards. The P for aces is then: P = 4/52 or 1/13. The same is true for drawing a king: P = 4/52 or 1/13. Using the *addition rule* for finding the probability of drawing a king *or* an ace we get: P = 1/13 + 1/13 = 2/13.

The addition rule can be extended to cover more than two events. Thus the probability for success of drawing A or B or C or . . . K for any one drawing or trial can be found if:

A. A, B, C, . . . K are mutually exclusive events.
B. The probabilities for drawing A, B, C, . . . K have the same denominator.
C. A, B, C, . . . K can occur at the same time or simultaneously in the same experiment.

(A, *go to* 28)
(B, *go to* 29)
(C, *go to* 30)

---

28 A, B, C, . . . K must be mutually exclusive. You are correct.

(*Go to* 31)

---

29 If A, B, C, . . . K do not have the same denominators for their separate probabilities, these denominators can easily be changed so that all probabilities in the series have the same lowest common denominator.

The correct answer is that A, B, C, . . . K must be mutually exclusive. To say that A, B, C, etc. are mutually exclusive means that A and B and C cannot possibly occur simultaneously in the same experiment. Therefore, in the example of the probability for thowing a 2 or a 5 in one throw of the die, it would be impossible to throw a 2 *and* 5 in the same throw. Thus the events or successes would be mutually exclusive.

(*Go to* 31)

---

30 Actually the provision for the addition rule is just the opposite of statement C. A, B, C, . . . K *cannot* occur at the same time or simultaneously in the experiment. The probability for successes must be mutually exclusive. In the example of the probability for throwing a 2 *or* a 5 in one throw of a die, it would be impossible to throw a 2 *and* a 5 in the same throw. Therefore the events or successes would be mutually exclusive.

(*Go to* 31)

---

31 The preceding calculation for finding probability was called the *addition rule*. It is in answer to the question of P(A *or* B). Which of the following examples would illustrate the need for calculating probability using the addition rule?

A. The probability for *not* drawing an ace or king in any one attempt from a deck of 52 cards
B. The probability for throwing a 1 and a 6 in one throw of two dice
C. The probability for throwing a 6 and a 1 in one throw of a die

(A, *go to* 32)
(B, *go to* 33)
(C, *go to* 34)

---

32 Choice A is the correct answer. The events are mutually exclusive, and you are asked the probability for not drawing any card but an ace *or* a king. To calculate the probability, first calculate the probability for drawing an ace or king (1/13 + 1/13 = 2/13). To find the probability for not drawing an ace or king, subtract the probability of getting a king or ace from 1 or [1 – (2/13) = 11/13]. This will give the ratio for the probability of not drawing the king or the ace. The reason that you subtracted from 1 was that the figure 1 represents certainty or the maximum probability statement.

(*Go to* 35)

---

33 You said that the addition rule would apply to the calculation of the probability for throwing a 1 *and* a 6 in one throw of *two* dice. You are not correct. The events are not mutually exclusive because there is the possibility of obtaining both successes in one throw of two dice. A property of the addition rule is that it determines the probability for one kind of success *or* another. The statement you selected asks for two kinds of success to occur during one throw of two dice. The addition rule demands that events be mutually exclusive.

(*Go to* 32)

---

34 The probability of throwing a 6 *and* a 1 in one throw of a single die would not illustrate a need for the addition rule because it would be impossible for both events to occur in one throw of the die. The statement requires that the events not be mutually exclusive and that both be obtained in one throw of the die. Which of the remaining choices would most adequately represent the need for the addition rule?

Continued.

A. The probability for *not* drawing an ace or king in any one attempt from a deck of 52 cards
B. The probability for throwing a 1 and a 6 in one throw of two dice

(A, *go to* 32)
(B, *go to* 33)

---

35 The addition rule applies only to mutually exclusive events. However, it is often desirable to know probabilities for joint events that are not mutually exclusive. For instance, let us say that A represents the probability of choosing aces in a deck of cards and B represents the probability of choosing hearts or diamonds.

What would the joint (A and B) probability represent?

A. Probability for drawing an ace of diamonds
B. Probability for drawing aces of diamonds or hearts

(A, *go to* 37)
(B, *go to* 36)

---

36 The probability of A and B [P(A and B)] represents the probability for drawing aces of diamonds or aces of hearts. You are correct. The ratio is 1/26 and computed in the following manner:

Probability for drawing aces = 4/52

Probability for drawing hearts (13/52) or diamonds (13/52) = 13/52 + 13/52 = 26/52 = 1/2

P(A and B) = 4/52 × 1/2 = 4/104 × 1/26

There is one chance in twenty-six that an ace of hearts or an ace of diamonds will be drawn.

(*Go to* 38)

---

37 You chose an incorrect answer. The joint probabilities of A and B would be represented by the probabilities of drawing aces of diamonds or aces of hearts. Its ratio would be 1/26. This is computed in the following manner:

Probability for drawing aces = 4/52

Probability for drawing hearts (13/52) or diamonds (13/52) = 13/52 + 13/52 = 26/52 = 1/2

P(A and B) = 4/52 × 1/2 = 4/104 = 1/26

There is one chance in twenty-six that an ace of hearts or an ace of diamonds will be drawn.

(*Go to* 38)

---

38 The product of the probabilities is yet another property of probability theory that enables one to calculate the chances for getting both A *and* B events when A and B events are *independent*. The characteristic of independence means that the probability of event A occurring remains the same regardless of whether B has occurred, and vice versa. Therefore, in the example previously given, the color and face of the card are independent.

The *multiplication rule*, as the previously stated relationship is called, is most easily illustrated in the replication of experiments in which the independence of events is assured. For instance, flipping a coin is one experiment. Flipping the coin ten times is a duplication of the experiment ten times with each experiment independent of the last. Which of the below would be the probability of flipping ten successive heads on ten flips of the coin?

A. $P = 1/2 \times 10 = 5.0$
B. $P = 1/10 \times 1/10 = 1/100$
C. $P = 1/2 \times 1/2 \times 1/2 \ldots 1/2 = (1/2)^{10} = 1/1024$

(A, *go to* 39)
(B, *go to* 40)
(C, *go to* 41)

---

39 You chose $P = 1/2 \times 10 = 5.0$. This should look wrong to you because the probability is equal to a whole number greater than one rather than a fraction. A probability is always expressed in values from 0 to 1. The correct answer is $P = \frac{1}{1024}$. When the multiplication rule is applied to independent events, the probability is calculated by multiplying together the probabilities for each independent event. Since the probability of obtaining heads for each experiment is 1/2, the probability for flipping ten successive heads is $(1/2)^{10}$ or $\frac{1}{1024}$.

(*Go to* 42)

---

40 $P = 1/10 \times 1/10 = 1/100$ is not the probability ratio for flipping ten successive heads. When the multiplication rule is applied to independent events, the probability is calculated by multiplying together the probabilities for each independent event. Since the probability of obtaining heads for each flip is 1/2, the probability for flipping 10 successive heads is $(1/2)^{10}$ or $\frac{1}{1024}$.

(*Go to* 42)

---

41 When the multiplication rule is applied to independent events, the probability is calculated by multiplying together the probabilities for each independent event. You are correct. Since the probability of obtaining heads for each flip is 1/2, the probability for flipping ten successive heads is $(1/2)^{10}$ or $\frac{1}{1024}$.

(*Go to* 42)

42 What would be the probability of rolling a 1 *and* a 6 (in that order) in two consecutive rolls of a die?

A. 2/6 or 1/3
B. 1/36
C. 1/9

(A, *go to* 43)
(B, *go to* 44)
(C, *go to* 45)

43 The answer 1/3 is incorrect. If the question were: What is the probability of rolling a 1 *or* 6 in *one* roll of a die, 1/3 would be the correct probability. However, you were to calculate the probability for rolling a 1 *and* a 6 in *two* rolls. These are independent events. The multiplication rule would apply in this instance.

The probability of getting a 1 in one roll of a die is 1/6, and the probability of getting a 6 in a second independent roll of the die is also 1/6. Therefore the probability of rolling a 1 *and* a 6 (in that order) in two consecutive rolls of a fair die = $P(1/6 \times 1/6) = 1/36$.

(*Go to* 46)

44 The correct probability of getting a 1 *and* a 6 (in that order) in two consecutive rolls of a fair die is 1/36.

The probability of getting a 1 in one roll of a die is 1/6, and the probability of getting a 6 in a second independent roll of the die is also 1/6. Therefore the probability of rolling a 1 *and* a 6 (in that order) in two consecutive rolls of a fair die = $P(1/6 \times 1/6) = 1/36$.

(*Go to* 46)

45 Your answer, 1/9, is incorrect. The probability of getting a 1 in one roll of a die is 1/6 and the probability of getting a 6 in a second independent roll of the die is also 1/6. Therefore the probability of rolling a 1 *and* a 6 (in that order) in two consecutive rolls of a fair die = P(1/6 × 1/6) = 1/36.

(*Go to* 46)

---

46 Could the multiplication rule be used for finding the probability of the chances for getting two aces in two draws from an ordinary deck of cards?

A. Yes
B. No
C. There is not enough information given to make a decision of whether it meets the criteria for using the multiplication rule

(A, *go to* 47)
(B, *go to* 48)
(C, *go to* 49)

---

47 It would be possible, but only if it were known whether the first card was replaced into the deck before the second drawing was made. If it was not replaced, the two events (two drawings) would not be independent. If the cards were drawn with replacement, the two draws would be independent, since the probability of getting an ace remains constant from one draw to the next. The result of the first cannot possibly affect the second.

(*Go to* 50)

---

48 It would be possible, but only if it were known whether the first card was replaced into the deck before the second drawing was made. If it was not replaced, the two events (two drawings) would not be independent. If the cards were drawn with replacement, the two draws would be independent, since the probability of getting an ace remains constant from one draw to the next. The result of the first drawing cannot possibly affect the second.

(*Go to* 50)

---

49 You are correct that there is not enough information given because the rule could be used only if it were known whether the first card were replaced into the deck before the second drawing was made. If it was not replaced, the two events (two drawings) would not be independent.

(*Go to* 50)

---

50 Which statement below illustrates a situation in which independence of events does *not* exist unless replacement is provided?

A. The probability of rolling a 2 and a 4 in four rolls of a die.
B. The probability of choosing a 9 in a single draw from a hat filled with numbers from 1 to 50, inclusive.
C. The probability of obtaining six single persons in a sample of ten persons drawn from a population of 100 adults.

(A, *go to* 51)
(B, *go to* 52)
(C, *go to* 53)

---

51 Independence of events does exist when rolling dice. The last throw does not depend on any previous throw. The correct answer is C. After making each selection, the total number of available persons decreases, and the probability changes unless the person from the previous selection is replaced back into the population and is given as equal a chance of being picked as the other members in the population.

(*Go to* 54)

---

52 There is only one outcome to be undertaken when making a single draw from the hat. Therefore there would never be an occasion when replacement is necessary. If two or more numbers were to be chosen, replacement would be necessary for the same reason that the correct answer, C, would have to have replacement. In choice C, after making each selection, the total number of persons available decreases, and the probability changes unless the person from each previous selection is replaced back into the population. With replacement, each person in the population has the same chance of being drawn as every other person on each draw. The probability of drawing a single person then would not depend on what kind of person was drawn on the previous selections.

(*Go to* 54)

---

53 It is correct that independence does not exist in sampling from a small population unless there is replacement. After making each selection, the total number of persons available decreases, and the probability of selecting another single person changes unless the person from the previous selection is replaced back into the population. With replacement, each person in the population has the same chance of being drawn as every other person on each draw. The probability of drawing a single person then would not depend on what kind of person was drawn on the previous selections.

(*Go to* 54)

---

54 There are times when a probability situation will require incorporating both the addition and multiplication rule to calculate the probability ratio. For instance, in calculating the chances for rolling a 1 and a 6 *in any order* with two rolls of a die, the situation is different than any previous example because the *order* of events for successes can be in any possible combination. Order of events refers to whether you roll a 6 first and then a 1, or a 1 first and then a 6. In this case there would be only two possible *combinations* or order of events that could occur.

If your task is to find the probability of getting an ace, king, and queen in three draws from an ordinary deck, with replacement, how many successful combinations of sequences or orders could you possibly obtain?

Three, six, or nine?

(*Go to* 55)

---

55 There are six possible combinations:

AKQ	KQA
AQK	QAK
KAQ	QKA

To calculate the probability of getting an ace, king, *and* queen in three draws from an ordinary deck, with replacement, the probabilities for each combination (AKQ, AQK, etc.) must ultimately be summed. Thus with each combination, the probability would be:

$$(4/52 \times 4/52 \times 4/52) \text{ or } (1/13)^3$$

Adding the combinations of probabilities together, the result would be:

$$6(1/13)^3 \text{ or } 6/2197$$

Calculating probability for *any order* as opposed to a specific order of drawing a series of events does what to the likelihood that success will occur?

A. Increases likelihood (A, *go to* 56)

B. Decreases likelihood (B, *go to* 57)

---

56 When any order of events is accepted, the likelihood that success will occur is increased compared to that likelihood when only a specific order of events is accepted.

The probability for one particular order or combination of drawing ace, king, and queen is $(1/13)^3 = 1/2197$, or one chance in every 2197 attempts. For *any* combination the likelihood becomes $6(1/13)^3 = 6/2197$, or six chances in every 2197 attempts, which represents a substantial increase in likelihood.

(*Go to* 58)

57 Actually the likelihood of drawing an ace, king, and queen in three draws is increased when any order is accepted as a success. If only a particular order is desired, the probability would be $(1/13)^3 = 1/2197$, or one chance in 2197 attempts. In the case of accepting any of the combinations as successful events, the ratio is $6(1/13)^3 = 6/2197$, or six chances in every 2197 attempts.

(*Go to* 58)

58 What is the probability of rolling a 1, 2, and 5, *in that order,* from three rolls of a die?

A. $(1/6)^3$ or 1/216
B. $6(1/6)^3$ or 1/36
C. $(1/2)^3$ or 1/8
D. Impossible to calculate because information concerning indepedence is insufficient

(A, *go to* 59) (C, *go to* 61)
(B, *go to* 60) (D, *go to* 62)

59 The answer is 1/216 as you stated. The probability for rolling a 1 is 1/6; for a 2, 1/6; and for a 5, 1/6.

Applying the multiplication rule, $(1/6)^3 = 1/216$.

Independence of events can be recognized by the nature of the event. Each event or roll is separate and not dependent on a previous roll.

(*Go to* 63)

60 The answer you gave (1/36) is not correct. The correct answer is 1/216.

You were asked for the probability of rolling a 1, 2, and 5, *in that order,* from three rolls of a die. The multiplication rule is applied to the separate probabilities ($1/6 \times 1/6 \times 1/6 = 1/216$). Since only one combination is called for, the addition rule does not apply.

Independence of events can be recognized by the nature of the event. Each event or roll is separate and not dependent on the outcome of a previous roll.

(*Go to* 63)

61 The answer you should have arrived at is 1/216. The probability for getting a 1 in the first roll is 1/6, for getting a 2 on the second roll it is 1/6, and for getting a 5 on the last roll it is 1/6 also. Applying the multiplication rule, the probability for this particular order is $(1/6)^3$ or 1/216.

Independence of events can be recognized by the nature of the event. Each event or roll is separate and not dependent on the outcome of a previous roll.

(*Go to* 63)

---

62 Independence of events, in this case, can be recognized by the nature of the event itself. Each event, or roll, is separate and success is not dependent on a previous roll. That is, what occurs in the first roll has no effect on what occurs in subsequent rolls. Therefore the second roll is independent of the results of the first or third rolls.

The correct answer is 1/216. This is obtained by multiplying the probability ratios of each throw of the die ($1/6 \times 1/6 \times 1/6 = 1/216$).

(*Go to* 63)

---

63 If events are not independent, how is probability affected? When it is recognized that the probability of A may be dependent on whether B occurs, it is called *conditional probability*. Thus, if B is the event that a man drinks excessively and A the event that he develops cirrhosis of the liver, then we would expect the probability of A (cirrhosis) given that B (drinking) has occurred to be greater than the probability of A without B, since excessive drinking is a cause of cirrhosis. The example involves a situation in which two variables are related so that knowledge of one helps to predict the other.

Independence of events does not exist in cases in which replacement is not maintained. For example, getting two aces in two draws from an ordinary deck of cards can be two events whose probabilities are not independent. If A is the event that one gets an ace on the first draw and B the event that an ace turns up on the second draw, A and B events are not independent unless the card first picked is reshuffled into the deck before the second selection is made. The probability for drawing an ace in the first attempt would be 4/52. If there is an ace drawn on the first attempt and it is not returned, what would the probability be for drawing another ace on the second draw?

A. 4/52
B. 3/51
C. 4/51

(A, *go to* 64)
(B, *go to* 65)
(C, *go to* 66)

---

64 Your answer, 4/52, is not the probability on the second draw. Since the card was not returned from the first drawing, only 51 cards remain. Also, since the ace was drawn on the first draw, there could only be three aces left to draw; therefore the probability of getting a second ace would be P = 3/51.

(*Go to* 67)

65 The probability of drawing an ace on the second draw is 3/51, since only 3 aces remain in the deck and there is now a total of only 51 cards.

(*Go to* 67)

66 Your answer, 4/51, assumes first that there are still 4 aces left in the deck, but it was stated that the first card selected was an ace; therefore there are only 3 aces left in the deck. You correctly chose 51 as the correct denominator, since there are now 51 cards in the deck because you already drew one. Therefore, if an ace is drawn on the first attempt and not returned to the deck, the probability of selecting a second ace will be 3/51.

(*Go to* 67)

67 If an ace was not selected on the first draw, and if the card that was selected was not returned to the deck, then the probability for getting an ace on the second draw would be 4/51. The computation of the probability of getting 2 aces in 2 draws needs to be computed as a *conditional probability* that stipulates that:

The probability of getting both A and B is the product of the probability of getting one of these events times the conditional probability of getting the other, given that the first event has occurred.

Thus if the probability for the example of drawing two aces in two draws from a deck is calculated without replacement, it is assumed that an ace was drawn on the first attempt. The calculated probability would be:

A. P(A and B) = 4/52 × 4/52 = 1/169
B. P(A and B) = 4/52 + 3/51 = 30/221
C. P(A and B) = 4/52 × 3/51 = 1/221

(A, *go to* 68)
(B, *go to* 69)
(C, *go to* 70)

68 You selected 4/52 × 4/52 = 1/169.

The first probability statement (4/52) is that of drawing an ace on the first draw and is correct, but the second probability statement should be 3/51 and represents the probability of drawing an ace on the second draw *given* that the first draw also produced an ace. Therefore the probability of drawing two aces from a deck without replacement would be:

$$P(\text{A and B}) = 4/52 \times 3/51 = 1/221$$

The correct answer is 1/221.

(*Go to* 71)

69 Your answer, 30/221, is not the correct probability for drawing two aces without replacement. The probabilities for each drawing are correct, but the multiplication rule must be applied, since we want the conditional probability of first one ace *and* then another ace, rather than the probability for one ace *or* another ace. The answer should be:

$$P(\text{A and B}) = 4/52 \times 3/51 = 1/221$$

(*Go to* 71)

70 The probability of drawing two aces in two draws from a deck is:

$$P(\text{A and B}) = 4/52 \times 3/51 = 1/221$$

(*Go to* 71)

71 The following points summarize probability as discussed in this text:

1. Probability refers to what will happen in the long run.

2. The probability for any given condition or experiment is the ratio of the number events or outcomes of interest to the total possible events or outcomes, expressed in values from 0 to 1.

$$P = \frac{f}{n} = \frac{\text{Number of outcomes of interest}}{\text{Total number of possible outcomes}}$$

3. If one wants to know the probability for one *or* another event or outcome occurring, it is determined by adding the probability ratios for each experiment. The *addition rule*, as previously stated, demands that events A or B be *mutually exclusive*; that is, A and B cannot occur simultaneously in the same experiment.

4. If one wants to know the probability for having A *and* B occurring, it can be done by multiplying the separate probabilities for A and B. The *multiplication rule*, as just stated, demands that A and B be *independent*. This means that the probability of event A occurring must remain the

Continued.

same regardless of whether B has occurred, and the probability of B is not at all affected by the event A.

5. Calculating probability for *any order* as opposed to a specific order of events will increase the likelihood that success will occur; that is, if any order is accepted, then all *combinations* of successful events must be considered.

6. When events are not independent and it is recognized that the probability of A may be dependent on whether B occurs, the probability ratio is called a *conditional probability*.

(*End of chapter* 6)

chapter

7

Statistical inference and methodology

At the conclusion of this chapter, you should be able to:

1. State the idea that science involves making decisions or drawing conclusions under conditions of uncertainty by making inferences or expectations (probability statements) based on prior experience (data) or theory.
2. Distinguish between cases in which hypotheses are arrived at by deduction or by induction.
3. Define and distinguish between the concepts of internal validity and external validity in research design.
4. Describe the purpose of randomization as a technique in research methodology.
5. Produce acceptable one-sentence definitions of a random sample and a simple random sample. (A sample is random when each case in the population has the same probability or a known probability of being selected; a sample is simple random when not only each case but each combination of cases has an equal chance of selection.)
6. Correctly label illustrative samples as being random or nonrandom.
7. Given a research study design or hypothesis, identify the variables as independent, dependent, control, or intervening.
8. Define and distinguish between the concepts of validity and reliability in measurement.

---

1 When one makes generalizations (e.g., generalizing from a sample to a population), predictions, estimations, or otherwise arrives at decisions in the face of uncertainty, the process illustrates what is done in:

A. Descriptive statistics (A, *go to* 2)

B. Inductive statistics (B, *go to* 3)

---

2 Your answer, descriptive statistics, is incorrect. *Descriptive statistics* is any treatment of numerical data that does *not* involve making generalizations from a sample to a population. The correct answer, inductive statistics, depends on *statistical inference* for making generalizations.

(*Go to* 4)

---

3 *Inductive statistics,* indeed, uses the process of *statistical inference,* which enables decisions or conclusions to be made about a population on the basis of generalizing from a sample of that population. (*Go to* 4)

---

4 We have said that statistics is the study of the methods of making decisions under conditions of uncertainty. That is what science is about —making decisions or drawing conclusions by using inferences from past experience. Thus we can view the scientific world as having the discipline of statistics and the parastatistical disciplines such as medicine, psychology, and biology. All sciences are based on inferences and probabilistic statements.

Fundamental to statistical inference is the fact that in research, as in all of science, we can never know the truth about something. All we can do is make an approximation or estimation or expectation. Mark Twain, in his *Life on the Mississippi* (1883), offered the following comments on the hazards of inference:

> In the space of one hundred and seventy-six years the lower Mississippi has shortened itself two hundred and forty-two miles. That is an average of a trifle over one mile and a third per year. Therefore, any calm person, who is not blind or idiotic, can see that in the Old Öolitic Silurian Period, just a million years ago next November, the Lower Mississippi River was upward of one million three hundred thousand miles long, and stuck out over the Gulf of Mexico like a fishing-rod. And by the same token any person can see that seven hundred and forty-two years from now the Lower Mississippi will be only a mile and three quarters long, and Cairo and New Orleans will have joined their streets together, and be plodding comfortably along under a single mayor and a mutual board of aldermen. There is something fascinating about science. One gets such wholesale returns of conjecture out of such a trifling investment of fact.

(*Go to* 5)

---

5 Statistics enable us to judge the *probability* that our inferences or estimates are close to the truth. It is the purpose of research design, including sampling and random assignment of cases, to control conditions so that our estimates and statements are more accurate and so that we can say just how accurate we are. Sometimes we are willing to assume conditions we did not control, such as normality or randomness.

Statistics, thus, is a method of reducing uncertainty and so allowing us to make decisions or draw conclusions. We reduce uncertainty in part by using a few descriptive statistics to represent unmanageable amounts of raw data. We also reduce uncertainty by letting our past experience or theory guide us in setting expectations about some event and then comparing what we observe with what we expect. A third way of reducing uncertainty is by following procedures in which we have confidence as to their leading to information useful and valid for conclusion-drawing or decision-making. This chapter deals briefly with the latter information—the concepts and methods of the scientific process, often called *methodology*, by which we obtain data for statistical analysis.

Understanding the scientific literature in your field is more a problem of understanding the scientific method and potential sources of error than of understanding statistics. Statistics can be helpful only when the method is sound.

(*Go to* 6)

---

6 The basic method of science is to propose alternative explanations of some problem of interest and then to seek ways to reject or test each of the alternatives. Those alternatives that cannot be rejected remain as the most plausible explanations. The remaining explanations are not proved; they are merely accepted as possible explanations. This method is called *negative inference*. Sometimes it is possible to set up a critical test among alternative explanations so that if one outcome occurs, alternative A must be right and alternative B must be wrong, whereas if another outcome occurs, then B must be right and A must be wrong. This approach, sometimes called *strong inference*, has enabled the field of biology to make tremendous advances in only a few years.

Alternatives may be stated as general statements integrating causes and effects, that is, as theories; or alternatives may be statements of specific hypotheses. Alternatives are generated sometimes by looking at theories or concepts and deducing or hypothesizing likely outcomes in a specific situation *(deduction)*. Alternative explanations may also be generated by considering previous research and empirical results and generalizing to a hypothesis *(induction)*.

Continued.

In the following problem was the hypothesis arrived at by deduction or induction?

Cognitive balance theory suggests that behaviors chosen by persons will be more strongly supported by the person than behaviors required of the person by external authority.

Hypothesis: Eighth-grade students who select an individualized self-study program in biology will have a higher mean score on attitude toward the class at the end of the year than will students who were randomly assigned into the self-study class.

A. Deduction

B. Induction

(A, *go to* 7)
(B, *go to* 8)

---

7 The hypothesis was arrived at, as you said, by deducing a specific outcome as a prediction or application from the theory on cognitive balance.

(*Go to* 9)

---

8 The hypothesis was not arrived at by induction. Instead, it was deduced from the general theory on cognitive balance. The general statement was applied to or operationalized into a specific instance. The hypothesis could have been arrived at by induction if the researcher had observed student attitudes in previous years in self-study classes in American problems and formulated the hypothesis in frame 6 as being consistent with his earlier observations.

(*Go to* 9)

---

9 The research process by which we implement the scientific method has (or should have) several characteristics.

Research is logical. The scientific logic just described, methodology logic, and the probabilistic-inferential logic discussed in the rest of this text should be used to validate a research process.

Research is rigorous. Research is structured and systematic. The report of research should display the structure for others to judge. Care must be observed in the procedures followed. Assumptions should be known and reasonable, and conclusions must be related to actual findings.

Research is empirical. Because research is referenced to reality, data are among the results of research. Most of these data must be statistically analyzed. Not all logical decision-making involves empirical data, but research must be empirical. Furthermore, empirical research involves measurement of operationally defined variables and explicit hypotheses.

The researcher must convert each variable from a concept to an operational form that can be observed and measured.

Research is transmittable. Research is recorded. It should be recorded clearly enough that someone else can repeat it (replicate it). A report of research should display the logic, rigor, empirical operations, and results clearly enough that decision makers can evaluate the value of results for decisions, and other researchers can build on the results for further studies.

(*Go to* 10)

---

10 A final characteristic of research that must be worked for is *validity*. The research methodology must be sound enough that the results can be accepted as a reasonable approximations and applicable to other situations. Problems leading to invalid research will be discussed in the final chapter of this text. For now, however, consider two kinds of validity in research.

A study has *internal validity* if the outcome of the study is a function of the program or approach being tested rather than the result of other causes not dealt with in the study.

A study has *external validity* if the outcome obtained would apply in the real world to other similar programs, approaches, or situations.

Internal validity: Is there a result?
External validity: Can the result be generalized?

Does the following study involve a problem of internal validity or external validity?

A medical school plans to introduce a new psychiatric medicine teaching program. Second-year students are surveyed as to their professional identification with psychiatry just before the program is introduced and again just before graduation two years later. The research question is: Does the new program increase identification with or selection of psychiatric specialities?

A. Internal validity
B. External validity

(A, *go to* 12)
(B, *go to* 11)

---

11 The study does not necessarily raise questions of external validity—are the results generalizable? Only if one could argue that succeeding classes or classes in other schools were different from the class studied would you have a problem of external validity.

(*Go to* 12)

---

12 The design does raise questions of internal validity—was there a result? To show change related to program, the research hypothesis, we must rule out change related to other factors such as simply growing older or being in the building for two years.

A more thorough discussion of validity and sources of error is presented in chapter 12.

(*Go to* 13)

13 Research studies have been categorized in various ways. Some persons like to consider basic research versus applied research. Basic research refers to finding fundamental relationships linked to a model or theory without any necessary immediate application. Applied research tries to test a problem solution for immediate use. Better terms have been suggested such as *conclusion-oriented* research aimed at drawing conclusions about general outcomes and *decision-oriented* research aimed at providing specific findings for making a decision.

A useful distinction in kinds of research studies is that of *experiments* versus *investigations* (or surveys). Experiments attempt to control significant components of the problem, whereas investigations take the world as it is and look for meaningful associations or differences. Since studies often incorporate both approaches, the relevance of the labels is in focusing our attention on sources of error. The more tightly controlled the experiment is, the better the internal validity may become, but the setting becomes more artificial and generalizing to "real" situations becomes more difficult. On the other hand, investigations may be in settings from which it is reasonable to generalize, but they may offer so little control of design that many alternative explanations of results remain unchallenged.

(*Go to* 14)

14 Many other labels for kinds of studies are specific to certain fields or describe kinds of controls. *Clinical trials* are planned studies of effects of therapeutic materials or programs in medicine, clinical psychology, nursing, etc. *Bioassays* are experiments on biological material aimed at estimating the potency of an unknown agent relative to some existing standard. In *evaluative research*, which abounds in education, decision-oriented studies attempt to guide education policy makers. *Case studies* try to draw generalizations by inference from in-depth study of one or a few example cases. *Blind* and *double-blind* studies refer to design methods in clinical studies in which either the subject or both the subject and the person administering the treatment are kept uninformed of what treatment is being used or what outcome is expected. *Longitudinal studies* are studies carried out over a prolonged period of time rather than by collecting data at only one or two points in time. *Cross-cultural studies* simply study a similar problem in more than one culture.

(*Go to* 15)

---

15 The design of research is a plan the researcher will follow to collect the appropriate data and control as many sources of error as possible. Usually the researcher is interested in looking at one or more variables. A variable is some measure that varies. It is the *variation* in which we are interested. The researcher attempts to understand the variation due to the elements being studied and to control or eliminate variation due to extraneous matters or to errors.

One of the principal methods of controlling the effects of extraneous variables or forces is *randomization*. R. A. Fisher, one of statistical methodology's outstanding developers, formalized randomization between 1920 and 1940 as the basis for much of research methodology. When comparing two groups, one would like them to be identical except for the treatment being studied. Sometimes the researcher can match the two groups on important variables (e.g., age, education). But you cannot match on every variable, and there always remains the chance that the groups differ in some unknown way that may affect the result of the study. Randomization, randomly assigning cases to groups, for example, tries to distribute all other unknown variation evenly, by chance, without bias, to each group. Randomization does not guarantee identical groups; that is impossible. Randomization allows the statistician to expect only moderate differences that will tend to follow chance distributions. Failure to follow randomization, thus producing samples of unknown bias, is one of the leading sources of error in the research that you will read. Randomization can apply to selecting samples, experimental treatment, or to assigning or distributing such things as days for measurement, test forms, observers, or any design factor that may affect outcomes. You will most commonly encounter randomization in the selection of samples.

(*Go to* 16)

---

16 You should keep in mind that the use of both *probability theory* and the theory behind *random sampling* are crucial assumptions that you will be using each time that you proceed with a test of statistical significance when testing a hypothesis.

Random sampling is a procedure for selecting individuals (or any other unit) from a population such that each individual has an equal or known chance of appearing in the sample.

Random sampling that gives each individual and each *combination* of individuals an equal chance of being selected is known as *simple random sampling.*

From this knowledge of random sampling, would it be wise to use random sampling as a way of ensuring independence of events or cases in an experiment?

A. Wise (A, *go to* 17)
B. Unwise (B, *go to* 18)

---

17 Technically you are incorrect. Since practically all sampling is done without replacement, the assumption of independence cannot be met. However, when the population is relatively larger than the size of the sample, one can safely neglect the resulting minor distortion due to the inability of an individual being drawn a second time. For example, if 25 persons are selected out of a population of 10,000, the chances are slight that any one person being selected as the first person in that sample would again be selected even if he were replaced. So, practically speaking, random sampling is wise if the size of the sample is relatively smaller than the size of the population. (*Go to* 19)

---

18 Technically you are correct; it would be unwise under certain circumstances. Most statisticians, however, do use random sampling as a procedure for alleviating the problem of independence. Since practically all sampling is done without replacement, the assumption of independence cannot be met. However, whenever the population is relatively larger than the size of the sample, one can safely neglect the resulting minor distortion that results when an individual cannot be selected a second time. For example, if 25 persons are selected out of a population of 10,000, the chances are slight that any one person being selected as the first person in that sample will again be selected even if he were replaced. If the sample is relatively smaller than the population, random sampling is a wise choice for obtaining independence. (*Go to* 19)

---

19 One of the best ways to get a simple random sample is to use a table of random numbers. (These tables can be found in appendices of many statistics textbooks.) In this way an experimenter can be assured of giving everyone in the population an equal chance of appearing, and the experimenter also is:

A. Guaranteed equal results for all conditions being tested
B. Assured control over all variables in the experiment
C. Assured that each combination of individuals is given an equal chance of selection

(A, *go to* 20)

(B, *go to* 21)

(C, *go to* 22)

---

20 The results should not be influenced by random sampling procedures. Rather, random sampling should guarantee that the results are not biased by the subjects or people who were selected. In experimental designs the "control" of bias is the essence of using random sampling procedures.

The correct answer is that simple random sampling also assures the experimenter that each *combination* of individuals is given an equal chance of selection.

(*Go to* 23)

---

21 Simple random sampling does not assure control over all variables in the experiment. It does not, for instance, control the way an experiment is administered. It does control any bias in the way people are selected by (1) giving everyone an equal chance of being selected and (2) giving each *combination* of individuals an equal chance of being selected.

(*Go to* 23)

---

22 Simple random sampling not only guarantees that everyone in the population will have an equal chance of appearing in the sample, but it also guarantees that each *combination* of individuals is given an equal chance of being selected.

(*Go to* 23)

---

23 There are many kinds of random sampling. Simple random sampling is only one kind. A sample in which you systematically take every tenth person in a list after a random start is random (each person has an equal chance of being drawn). A sample that takes every male and every third female is random (each person has a *known* probability of being sampled), but it is not simple random. Many other kinds of samples you may read about (stratified, clustered) are also random.

If, however, the research reports that the sample was selected because of some interesting characteristic in those selected or because they were available (the man-in-the-street approach), then you can assume that some bias has entered and that the sample well may be different from the population to which you wish to generalize.

(*Go to* 24)

---

24 One of the elements of design, in addition to sampling, is the identification of the variables of interest. We have already mentioned one consideration, that of *operational definitions* for variables. The vague concept intelligence must be operationally defined as the score on an IQ test before you can actually accumulate data.

(*Go to* 25)

---

25 Variables also are labelled according to their role in a study such as independent variables, dependent variables, and intervening variables. Many different labels may be found in the literature, but for now consider only the following categories:

Independent variables: Those stimulus, treatment, or input variables measured or manipulated by the researcher to see what happens

Dependent variables: Those response, criterion, or output variables measured (but not manipulated) to see what effect has occurred

Control variables: Those variables identified and controlled in the study so as to neutralize any affect they might have had on the results

Intervening variables: Those factors which cannot be measured or manipulated but which are inferred to be acting on the observed dependent variables as a result of the action of the independent variable; that is, they intervene between independent and dependent variables

In the following hypothesis, which is the dependent variable?

Among pre-school age children, accuracy in identification of left and right is related to the amount of practice in left-right movement games.

A. Age
B. Left-right identification
C. Learning
D. Amount of practice in movement games

(A, *go to* 26) (C, *go to* 28)
(B, *go to* 27) (D, *go to* 29)

---

26 The dependent variable is the response or criterion used to measure effect. Age is not the variable stated as the result in which we are interested. Go back and try again.

(*Go to* 25)

---

27 Left-right identification is the dependent variable, the criterion we measure to study the outcome.

(*Go to* 30)

---

28 Learning was not explicitly mentioned. Remember that the dependent variable is the response or criterion we measure to determine results. Go back and try to decide which variable must be measured to determine outcome.

(*Go to* 25)

---

29 The amount of practice in left-right movement games is not the dependent variable. It precedes and is hypothesized to lead to left-right identification. The dependent variable is the response or criterion we measure to determine results. Go back and try to decide what the outcome variable is.

(*Go to* 25)

---

30 Using the same example on left-right learning, match each of the variables with the appropriate label.

1. Age	____________
2. Left-right identification	Dependent variable
3. Learning	____________
4. Amount of practice in movement games	____________

(Labels: Dependent, Independent, Control, Intervening)

(*Go to* 31)

---

31 The variables are as follows:

1.	Age	Control variable
2.	Left-right identification	Dependent variable
3.	Learning	Intervening variable
4.	Amount of practice in movement games	Independent variable

The basic fact to remember is that the independent variables are those which do the acting, and the dependent variables are those which are acted on.

(*Go to* 32)

32 Because research is empirical, measurement of data (and, subsequently, statistical analysis of the data) is central to research. Measurement was considered briefly in chapter 1 when scales of measurement (nominal, ordinal, interval) were discussed. Two other characteristics of measurement should be learned here: reliability and validity.

Validity in measurement refers to whether what is claimed to be measured is in fact measured. Thus validity is how well a measure does what it says it does. For example, the opinions in letters to the President may not be a valid measure of public opinion because only some kinds of people (a subset of the public) write letters to the President.

Reliability in measurement refers to how consistently a measure measures whatever it is measuring. If repeated measuring of some variable keeps giving you the same results, the measure is reliable.

The two terms can be considered separately. A measure may be reliable but invalid. Research that is reporting a new way to measure some concept should indicate why the researcher believes the measure to be reliable and valid.

(*Go to* 33)

33 At the heart of research methodology is experimental design. Much research in the applied field or the professions is what we earlier called investigations or surveys. However, our basic sciences often conduct experiments, and even field studies often involve combinations of experimental and survey approaches in some quasi-experimental design. Four categories of designs will be briefly mentioned here: experimental, factorial, quasi-experimental, and ex post facto.

(*Go to* 34)

34 *Experimental designs* are those in which all factors in the design are controlled by the experimentor: samples, when and to whom treatments are applied, and when and to whom measurements are applied. Furthermore, experiments involve comparisons, so that some kind of control group is necessary in an experiment. At a minimum, the classical experimental design is to randomly assign a sample of subjects to one of two groups. Give each group a pretest, give one group the experimental treatments, then give each group a posttest. Assuming that the groups are alike on the pretest, differences shown by the posttest are ascribed to the treatment. Actually, in many cases no pretest is needed because of the powerful effect of randomization. Because of the need to eliminate other explanations of results, studies of effects of treatments must have some control group that does not receive the treatment. (For a detailed discussion see Cambell and Stanley, 1963.)

(Go to 35)

---

35 *Factorial designs* permit the simultaneous assessment of the effects of several factors. The classical experimental logic is to hold everything constant but one factor and then study the effects of that one factor. This approach, however, is both costly and tedious. Factorial designs (you may have heard terms such as *randomized blocks* or *Latin squares*) and advanced statistical techniques allow realistic evaluation of several factors simultaneously, much as these factors occur in natural settings. In addition, multi-factor studies allow the researcher to look at *interactions* between variables. Interactions, for example, occur when one group (e.g., boys) scores high in treatment A and low in treatment B, whereas another group (girls) scores low on treatment A and high on treatment B. There is an interaction between sex and treatment on the criterion score.

(*Go to* 36)

---

36 *Quasi-experimental designs* are designs in which some important aspect of experimental control is not possible. For example, the researcher may be able to control randomization of samples and the measurement process but not when the experimental stimuli is encountered. Quasi-experimental designs abound in the applied fields. When well done, they enable the researcher to reject alternative explanations of results, which is the very purpose of research design.

(*Go to* 37)

37 *Ex post facto designs* are those that attempt to create controls after the fact. For instance, an ex post facto design might take people with a certain characteristic (membership in Ku Klux Klan or death caused by lung cancer) and try to match them on various other characteristics with other persons without the characteristic under study, and then look at differences between the two samples on some intervening variable of interest (e.g., smoking). No matter how you match, the two samples are not comparable, and some alternative explanations cannot be rejected. Ex post facto studies lead to suspect conclusions regarding cause and effect.

(*Go to* 38)

38 Equally important as the design are the methods used in carrying out the study and data analysis. Tests and questionnaires must be created and administered, and perhaps interviews must be made. Data can be biased at its collection, such as through leading questions in an interview. Analyses must relate statistics used to the kind of variables used (for instance, whether a measure is nominal or ordinal). Conclusions must distinguish statistical significance (e.g., nonchance difference) from importance or clinical significance. For instance, with large samples, small differences can be statistically significant (that is, they can be judged as real differences), but the difference may be so small as to be meaningless for decision making. Conclusions must be related to the findings. You, the reader, will have to judge whether an article makes conclusions that are unsubstantiated by the data reported.

This chapter has attempted to provide a brief overview of methodology and making statistical inferences. The chapter has provided some background, methodological concepts or principles, and a discussion of the elements of design. Methodology, you should remember, consists of procedures designed to help eliminate sources of error or other alternative explanations for some phenomenon so that you may try to reject the null hypothesis in favor of an explanatory hypothesis or have more confidence in an estimate or conclusion. In reading the literature in your field, you must be able to offer challenging questions as to the validity of the study. Most of these questions focus on sources of error in design or statistics. Chapter 12 deals more with possible errors and other questions that can be put to a research report.

(*End of chapter* 7)

chapter

8

Sampling distributions

At the conclusion of this chapter, you should be able to:

1. With notes, apply the binomial distribution formula:

$$P(r) = \binom{N}{r} \times (p^r q^{N-r})$$

   for each r given N to compute a binomial distribution table.
2. Estimate expected outcomes in a series of binomial trials based on the principle that the proportion of various sets of outcomes observed in a series of binomial trials will approach the theoretical proportions derived from the binomial distribution as the number of trials gets larger.
3. State the central limit theorem ("the result that for repeated random samples of size N drawn from a population with mean μ the distribution of sample means will itself be approximately normal with a mean μ regardless of the shape or form of the population as long as N is large").
4. State what is meant by the "standard error" of a statistic (standard deviation of the sampling distribution for the statistic; represents error in our estimate of the statistic due to sampling fluctuations).
5. Compute the standard error of the mean given $\overline{X}$, s, and N.

1 You will recall that a *distribution* is a curve representing the frequencies of scores or measures for each value on the X axis scale. A frequency distribution for values of some variable in a sample is constructed after the sample is collected and the scores tabulated. Such a distribution is called a *sample distribution*. In a research experiment or investigation you are interested in some statistic (outcome) about the study (e.g., a mean or correlation). If one wanted to project or predict a distribution of occurrences for each possible value or outcome of an experiment before the experiment was made, the best approach would be to:

A. Assign the same frequency of scores or measures to all values on the X axis
B. Calculate the probability of occurrences for each value on the X axis and plot it into a distribution

(A, *go to* 2)
(B, *go to* 3)

---

2 Your answer would not provide the best way of projecting or guessing the likelihood of measures occurring at certain possible values of an experiment. Assigning equal weights or frequencies of occurrences to each value on the X axis, although systematic, assumes that each value or outcome will have an equal chance for occurring. As will be shown later, this is not always the case. It would be better to calculate the probability of occurrences for each value of an experiment. The resulting distribution would then be a probability distribution of all the possible outcomes of any given experiment. This concept, when expanded to a sampling distribution, will shed more light on the role of probability in hypothesis testing.

(*Go to* 4)

---

3 Indeed, the best approach would be to calculate the *probabilities* of occurrence for values of an experiment. The resulting distribution would then be a probability distribution of all the possible outcomes of any given experiment. This concept, when expanded to a sampling distribution, will shed more light on the role of probability in hypothesis testing.

(*Go to* 4)

4 A *sampling distribution* refers to the relative number of times one would expect to get certain outcomes in a large number of experiments. For example, if an experiment can either succeed or fail, what proportion of times do you expect success or failure? For another example, if you draw a sample of students and calculate mean age, what proportion of time do you expect a mean of 12.5 years, 12.6 years, or 13 years? In other words, a sampling distribution is a distribution of probabilities. The sampling distribution of probable outcomes for a great many events of interest is constructed through the use of the *binomial* expansion. Binomial refers to events that can occur in one of two ways. You may recall from high school algebra the binomial expression $(a + b)^n$. Examples of binomial events are experiments that either fail or succeed, yes-no voting, accept-reject inspections, presence or absence of traits, or right-wrong on beer testing. To illustrate the calculation of the binomial, the example of coin flipping will be used. If you recall, each time a coin is flipped it is considered to be:

A. A discrete experiment
B. Part of an experiment

(A, *go to* 5)
(B, *go to* 6)

---

5 A coin flip is a discrete experiment. Each flip of a coin is an independent event that will give a success or failure for one experiment. Several flips of the coin consist of several experiments, each experiment having the same probability for success or failure. One could also label the number of flips as being the sample size with interest given to the number of heads (successes) obtained in N flips.

(*Go to* 7)

---

6 A coin flip should not be considered a part of an experiment, but as a discrete experiment. Each time the coin is flipped, it is an independent trial that will show either success or failure for one experiment. Several flips of the coin consist of several experiments, each experiment having the same probability for success or failure. One could label the number of flips as being the sample size with interest given to the number of heads (successes) obtained in N flips. However, each flip would still be a separate experiment.

(*Go to* 7)

---

7 If asked to find the probability for getting six heads followed by four tails, you know that the total N must be:

A. N = 24 flips
B. N = 10 flips
C. N = 6 flips

(A, *go to* 8)
(B, *go to* 9)
(C, *go to* 10)

8 N does not equal twenty-four flips. N is the sample size or total number of experiments. If you want to flip six heads and 4 tails, it means you must do at least six flips for heads and four flips for tails, or a total of ten flips.

In this example, the number of successes is represented by the letter r, and the number of failures can be stated as N − r. Keep this notation in mind, since it will be used in the development of the binomial.

(*Go to* 11)

9 The total N is ten flips. In this example the number of successes is represented by the letter r, and the number of failures can be stated as N − r. Keep this notation in mind, since it will be used in the development of the binomial.

(*Go to* 11)

10 N does not equal six flips. It will take six flips to get six heads, but the sample size must also include four tails. This makes the total N equal to 10. In this example, the number of successes is represented by the letter r, and the number of failures can be stated as N − r. Keep this notation in mind as the binomial is further developed.

(*Go to* 11)

11 To obtain the probability of getting six heads followed by four tails, let p be the probability for obtaining heads (1/2). If q represents the probability of tails, then the likelihood for getting tails would be 1 − p = (1 − 1/2) = 1/2.

The probability of getting six heads followed by four tails, then, is the *product* of the *separate* probabilities for each experiment, or:

A. 1/2
B. 1/1024
C. 5/16

(A, *go to* 12)
(B, *go to* 13)
(C, *go to* 14)

12 Your answer, 1/2, is incorrect. The probability for each experiment (here, each flip of the coin) would be 1/2, but the probability of flipping six heads followed by four tails would be the product of the independent probabilities (1/2 for heads; 1 − 1/2 = 1/2 for tails). Then the probability ratio would be:

For heads: For tails:

$$(1/2 \times 1/2 \times 1/2 \times 1/2 \times 1/2 \times 1/2) \times (1/2 \times 1/2 \times 1/2 \times 1/2) = \frac{1}{2^{10}} = 1/1024$$

Or

$$p^r \times q^{N-r}$$

Where

p^r = the probability for getting six heads
q^{N-r} = the probability for getting four tails (*Go to* 15)

13 Your answer, 1/1024, is correct. The product of the probabilities for obtaining six heads (1/2 for each flip) followed by four tails (1 − 1/2 = 1/2 for tails) is calculated as shown below.

For heads: For tails:

$$(1/2 \times 1/2 \times 1/2 \times 1/2 \times 1/2 \times 1/2) \times (1/2 \times 1/2 \times 1/2 \times 1/2) = \frac{1}{2^{10}} = 1/1024$$

Or:

$$p^r \times q^{N-r}$$

Where

p^r = the probability for getting six heads
q^{N-r} = the probability for getting four tails (*Go to* 15)

14 Your answer, 5/16, is incorrect. You may have obtained this answer by *adding* the product of probabilities for heads (1/64) and for tails (1/16). This is incorrect. They should have been multiplied as shown below:

For heads: For tails:

$$(1/2 \times 1/2 \times 1/2 \times 1/2 \times 1/2 \times 1/2) \times (1/2 \times 1/2 \times 1/2 \times 1/2) = \frac{1}{2^{10}} = 1/1024$$

Or:

$$\begin{matrix} (1/64) & \times & (1/16) & = 1/1024 \\ p^r & \times & q^{N-r} & \end{matrix}$$

Where

p^r = the probability for six heads
q^{N-r} = the probability for four tails (*Go to* 15)

---

15 Thus far what actually has been done is to calculate the probability ratio for *one order* of a set of outcomes, or six heads *followed* by four tails (which is calculated by the formula $p^r q^{N-r}$). Another order for the general outcome of getting a total of six heads and four tails, other than six heads followed by four tails, would be to get two heads followed by four tails followed by four heads. No matter what order is desired for this set of outcomes (i.e., six heads and four tails) the same basic formula will apply. The probability ratio for all other ordering of this same set of outcomes would be:

A. Different
B. The same

(A, *go to* 16)
(B, *go to* 17)

---

16 The probability for all other orderings of the same outcome would not be different. If you apply the formula for finding the probability for any given set of outcomes ($p^r q^{N-r}$), you will note that the individual probabilities and their product remain the same for all ways that this particular set of outcomes (i.e., six heads and four tails) can be ordered. For example, if the order was two heads followed by four tails followed by four heads, the ratio would be equal to

$$(1/2 \times 1/2)\,(1/2 \times 1/2 \times 1/2 \times 1/2)\,(1/2 \times 1/2 \times 1/2 \times 1/2) = 1/1024$$

or the same as the probability for six heads followed by four tails.

(*Go to* 18)

---

17 Indeed, there would be no change in the probability ratio for any other outcome. The product of the probabilities would be the same. You can verify this by applying the formula ($p^r q^{N-r}$) to the other possible ways of ordering this set of outcomes: r = number of successes, N = total number of trials.

(*Go to* 18)

---

18 Each set of outcomes has a number of *ways it can be ordered*. Before the binomial distribution can be calculated, this factor of the total possible ways a set of outcomes can appear must be determined. Thus, in our example of six heads and four tails, one must ask how many ways can a coin be flipped ten times to get different orders of the same set of outcomes—six heads and four tails? That is, how many different *combinations* of six heads and four tails can you get out of ten flips of a coin?

When counting the ways, it usually becomes a difficult task because of the moderately large size of the sample (i.e., N = 10). Therefore a formula will again be used.

The formula for finding the number of ways a set of outcomes can be ordered is

$$\binom{N}{r} = \frac{N!}{r!\,(N-r)!} = \frac{N\,(N-1)\,(N-2)\ldots(N-r+1)}{r!}$$

where N = the size of the sample, or total number of flips, and r = the desired number of successes for the set of outcomes, or six heads.

The symbol r! is read "r factorial" and means that one should multiply r by r − 1, by r − 2, etc. until that reduction equals 1. $\binom{N}{r}$ is read as "the number of combinations of N things taken r at a time." What would be the number of ways the set of outcomes of six heads in ten flips of a fair coin could be ordered?

Calculate your answer, then . . .

(*Go to* 19)

---

19 The correct answer would be 210 ways.

The substitution in the formula and calculation would be as follows:

$$\binom{N}{r} = \frac{N\,(N-1)\,(N-2)\ldots(N-r+1)}{r!}$$

$$\binom{N}{r} = \binom{10}{6} = \frac{10 \times 9 \times 8 \times 7 \times 6 \times 5}{6 \times 5 \times 4 \times 3 \times 2 \times 1} = 210$$

Note that the number 5 is the last number to appear in the numerator because (N − r + 1) = (10 − 6 + 1 = 5). In plugging in the numbers in the numerator of the formula, it is suggested that you calculate the last figure (N − r + 1) first; then fill in the remaining numbers.

If the set of outcomes were two heads and eight tails, how many ways can this set of outcomes turn up in ten flips of a coin if you consider heads to be successes?

The formula is:

$$\binom{N}{r} = \frac{N\,(N-1)\,(N-2)\ldots(N-r+1)}{r!}$$

Calculate your answer, then . . .

(*Go to* 20)

---

20 N = 10, r = 2, and (N − r + 1) = 9. Therefore substituting into the formula,

$$\binom{N}{r} = \binom{10}{2} = \frac{10 \times 9}{2 \times 1} = 45$$

Now that the probability for a given outcome can be determined ($p^r q^{N-r}$), and the number of ways each set of outcomes can be ordered has been calculated $\binom{N}{r}$, the values of the binomial distribution, or the probabilities of *exactly* r successes, can be arrived at by multiplying

$$\binom{N}{r} \text{ by } p^r q^{N-r}$$

Therefore for the example of ten coin flips, how many *sets* of outcomes or probabilities of exactly r successes would have to be calculated for a complete binomial distribution? That is, how many different values could r take on within a ten-flip limit?

A. 10
B. 11
C. 100

(A, *go to* 21)
(B, *go to* 22)
(C, *go to* 23)

---

21 You must include the set of probabilities that would lead to zero heads. If you included zero times for getting heads, there would be eleven sets of outcomes. There could be:

0 heads and 10 tails
1 heads and 9 tails
2 heads and 8 tails
3 heads and 7 tails
4 heads and 6 tails
5 heads and 5 tails
6 heads and 4 tails
7 heads and 3 tails
8 heads and 2 tails
9 heads and 1 tails
10 heads and 0 tails
11 Total sets of outcomes (heads)

(*Go to* 24)

---

22 You are correct that r could take on eleven values (0 through 10) within a ten-flip limit.

(*Go to* 24)

---

23 You did not count correctly the number of possible values of r or successes that could occur within a ten-flip limit. If you counted the number of sets of outcomes, you would have eleven ways. There could be:

0 heads and	10 tails
1 heads and	9 tails
2 heads and	8 tails
3 heads and	7 tails
4 heads and	6 tails
5 heads and	5 tails
6 heads and	4 tails
7 heads and	3 tails
8 heads and	2 tails
9 heads and	1 tails
10 heads and	0 tails
11 Total sets of outcomes (heads)	

(*Go to* 24)

---

24 The binomial distribution for this example is, then, formulated by calculating the exact probability of each set of outcomes possible for the sample population. Here there are ten flips or eleven possibilities by using the formula:

$$P_r^N = \binom{N}{r} \times p^r q^{N-r}$$

P_r^N can be read as "the probability of r successes in N events." For example, the probability of two heads in ten flips is:

$$P_2^{10} = \binom{10}{2} \left(\frac{1}{2}\right)^2 \left(1 - \frac{1}{2}\right)^{10-2}$$

$$= \frac{10 \times 9}{2 \times 1} \left(\frac{1}{4}\right) \left(\frac{1}{256}\right) = \frac{45}{1024} = .044$$

Applying the binomial formula for each of the eleven possible outcomes would result in the distribution presented below:

Number of heads	*Probabilities (with P = 1/2)*
0	.001
1	.010
2	.044
3	.117
4	.205
5	.246
6	.205
7	.117
8	.044
9	.010
10	.001
	1.000

Continued.

It is obvious that the resulting distribution is a symmetrical distribution. The $\binom{N}{r}$ factor will always be symmetrical in:

$$P_r^N = \binom{N}{r} (p^r q^{N-r})$$

However, the $p^r q^{N-r}$ factor will prove symmetrical only when $p = q = ?$

A. 1/2
B. 1/4

(A, *go to* 25)
(B, *go to* 26)

---

25 1/2 is correct. Only with the probability of 1/2 for each event occurring will the final distribution give a perfectly symmetrical distribution.

(*Go to* 27)

---

26 You cannot define a term where $p = q = 1/4$, since $q = 1 - p$. Therefore the $p^r q^{N-r}$ factor will prove to be symmetrical only when $p = q = 1/2$.

(*Go to* 27)

---

27 The binomial distribution just illustrated was a simple example with only eleven sets of outcomes and with only two outcomes possible on each flip. Realistically, in most experiments, the possible outcomes will be large, and it may be necessary to *group* certain outcomes together and to associate a probability with the entire group of outcomes. Thus if a coin was flipped 1000 times, the probabilities might have been obtained for 400 to 449 heads, 450 to 499 heads, and 500 to 549 heads.

What does the binomial distribution tell us? If the coin flipping experiment is performed 1,024,000 times, and if you know that the distribution of all possible outcomes is represented as below, what would be the expected number of outcomes with no heads (all tails)?

Number of heads	
0	1/1024 = .001
1	10/1024 = .010
2	45/1024 = .044
3	120/1024 = .117
4	210/1024 = .205
5	252/1024 = .246
6	210/1024 = .205
7	120/1024 = .117
8	45/1024 = .044
9	10/1024 = .010
10	1/1024 = .001

A. 100
B. 45
C. 1000

(A, *go to* 28)
(B, *go to* 29)
(C, *go to* 30)

---

28 You should expect 1000 outcomes with no heads, not 100. If it is expected that *one* no-head outcome would occur for every 1024 times the procedure is tried, then in 1,024,000 times (1000 times 1024), 1000 no-heads would be expected.

As the number of performances of the experiment is increased, the observed empirical probabilities or proportions will get closer to these theoretical probabilities (i.e., .001).

(*Go to* 31)

---

29 You should expect 1000 outcomes with no heads, not 45. If it is expected that *one* no-head outcome would occur for every 1024 times the procedure is tried, then in 1,024,000 times (1000 times 1024), 1000 no-heads would be expected.

As the number of performances of the experiment is increased, the observed empirical probabilities or proportions will get closer to these theoretical probabilities (i.e., .001).

(*Go to* 31)

---

30 Yes, the binomial distribution will indicate an expected 1000 outcomes of no heads if the experiment is performed 1,024,000 times. As the number of performances of the experiment is increased, the observed empirical probabilities or proportions will get closer to these theoretical probabilities (i.e., .001).

(*Go to* 31)

---

31 A binomial distribution can be generalized to a sampling distribution when it is said that it:

A. Refers to the relative number of times we would expect certain outcomes
B. Refers to the relative number of times we would expect certain outcomes in a large number of experiments

(A, *go to* 32)
(B, *go to* 33)

---

32 You selected the wrong answer. It is important that you remember that a sampling distribution is a distribution of probabilities that refers to the relative number of times we would expect certain outcomes *in a large number of experiments.*

(*Go to* 34)

---

33 A sampling distribution, as you indicated, is a distribution of probabilities that refers to the relative number of times we would expect certain outcomes *in a large number of experiments.*

(*Go to* 34)

---

34 An important principle in statistics, relevant to the concept of a sampling distribution, is the *central limit theorem.* The central limit theorem states that the sampling distribution for some statistics will be approximately normal and of known values. For sample means, for instance, the central limit theorem states that if repeated random samples of size N are drawn from a population and if a mean is calculated for each sample, the distribution of sample means will approach the normal distribution as N becomes large, and the sampling distribution will have a mean approaching the population mean μ and a variance σ^2/N (σ^2 = variance of the population). This is true even if the population is not normal (e.g., a skewed distribution). The larger N is, the more normal the sampling distribution becomes and the closer the sampling distribution mean gets to the population mean.

(*Go to* 35)

---

35 A *sampling distribution* of a statistic (e.g., $\bar{X}$, $[\bar{X}_1 - \bar{X}_2]$, r, t) refers to the theoretical distribution of the statistic if you drew a large (infinite) number of samples of size N from a population, calculated the statistic for each sample, and plotted the distribution of all the calculated statistics.

No matter how unusual the distribution of the population when beginning, the sampling distribution for sample means can be counted on as being normal if:

A. N is sufficiently large
B. The samples that make up the distribution are normal

(A, *go to* 36)
(B, *go to* 37)

---

36 If N is sufficiently large, the sampling distribution for the mean will appear as a normal curve. N refers to the size of the sample.

The value of the central limit theorem is that it enables the researcher to relax concern over the question of whether the population is normally distributed. Knowing that the sampling distribution for means is normal, the researcher can proceed to use the normal curve in testing statistical hypotheses about the mean.

(*Go to* 38)

---

37 In tests of statistical significance about the mean, it does not matter if the samples that make up the distribution are normal or if the parent population is normal, *if* N is sufficiently large. A large N will result in a sampling distribution that is normally distributed.

The value of the central limit theorem is that it enables the researcher to relax concern over the question of whether the population is normally distributed. Knowing that the sampling distribution for means is normal, the researcher can proceed to use the normal curve in testing statistical hypotheses about the mean. (*Go to* 38)

---

38 Although no real population is perfectly normally distributed for a given characteristic, if the researcher were to draw numerous samples from that population, calculate the mean for each sample, and plot the means for each sample, a normal curve would result.

If in the population the characteristic has a mean of μ and a standard deviation of σ, then the distribution of sample means drawn from that population will have a mean of μ and a standard deviation of $\sigma/\sqrt{N}$. Remember that this is the standard deviation of the sample means. It is called the *standard error of the mean*, and is symbolically written as $\sigma_{\bar{X}} = \sigma/\sqrt{N}$. (*Go to* 39)

---

39 In a sampling distribution, each element in the distribution is an estimate of the population value. There is a certain amount of error in each of these estimates due to sampling fluctuations. The differences in estimates are the reason we have a distribution at all instead of just a single value. Because the differences represent errors in our estimate, we refer to the standard deviation of a sampling distribution as a *standard error*.

Given that the standard deviation of the sampling distribution for means (the standard error of the mean) is defined as $\sigma/\sqrt{N}$, then will the standard error become larger or smaller as the size of the sample is increased?

A. Larger (A, *go to* 41)
B. Smaller (B, *go to* 40)

---

40 The standard deviation of sample means or, more properly, the standard error of the mean for the sampling distribution of sample means will become smaller if the size of each sample is increased. (*Go to* 44)

FIG. 8-1

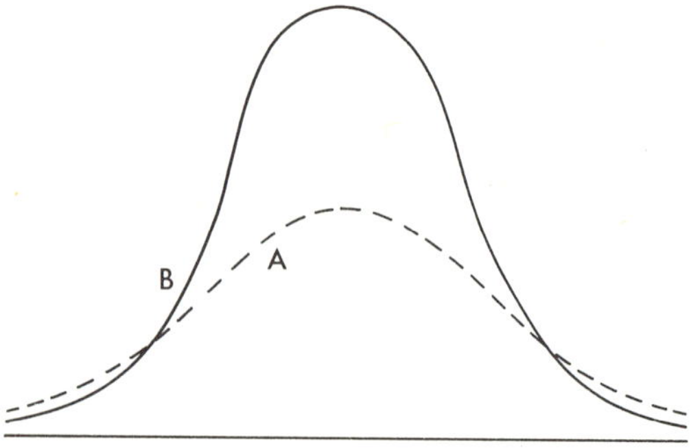

41 Consider the formula for the standard error of the mean: $\frac{\sigma}{\sqrt{N}}$. As N gets larger, the ratio must get smaller. The standard error of the mean (the standard deviation of the sampling distribution for means) will not become larger. As the sample size is increased, there is more clustering of means, which will result in a smaller standard deviation. Which curve in Fig. 8-1, then, would represent the sampling distribution of the mean with the larger sample size and thus the smaller standard error of the mean in the sampling distribution?

A. Curve A
B. Curve B

(A, *go to* 42)
(B, *go to* 43)

---

42 Distribution A is not the distribution based on the larger sample size. Curve A represents a larger standard deviation or standard error. The greater the clustering of the sample means the more peaked the distribution. Therefore distribution B represents a sampling distribution with a larger sample size. The sample means would be more highly clustered around the true population mean.

(*Go to* 44)

---

43 The sampling distribution with the smaller standard error is B. The sample means are more highly clustered around the true population mean.

(*Go to* 44)

---

44 Because the sampling distribution and not the population distribution is used in statistical tests of hypotheses, it is important to recognize the feature of clustering of means as the sample size is increased. It enables the researcher to have more faith in estimating the true population mean from a large sample than from a small one because, according to the central limit theorem, sample means will vary less from sample to sample if N is large.

(*Go to* 45)

---

45 Sampling distributions are essential in inferential logic because they are mathematically defined distributions, and thus we can know the relative frequency or probability of occurrence of any value in the distribution. In statistics we must compare the outcome of our experiment or investigation against some expected value to talk about significant results. A sampling distribution is a distribution of expected values. This chapter has introduced the sampling distribution through the binomial distribution. As the number of trials (N) in a binomial expansion becomes large, the binomial distribution approaches the normal probability distribution. A "normal" distribution is based on a curve whose equation is defined and which is bell-shaped, symmetrical, and unimodal. Because Karl Friedrich Gauss, the brilliant mathematician of the early 1800's, produced significant proofs about the normal curve (at age 10!), it is sometimes referred to as a Gaussian curve. Although all normal distributions are bell-shaped and symmetrical, not all bell-shaped and symmetrical curves fit the equation for a normal curve.

Statistical significance using normal distributions will be discussed at length in the next two chapters. There are a great many other distributions besides the binomial and normal that are used in statistics. Volumes have been written describing and tabling values of distributions such as t, chi-square, F, Poisson, and countless others. The Poisson distribution, for instance, is the distribution taken in many cases by events that happen only rarely in a large number of trials (e.g., being struck by lightning). Most statistical inferences you will encounter in the literature are based on normal distributions and a number of assumptions about the population parameter. Such statistics are referred to as *parametric* statistics. Some statistics, however, are based on inference procedures that rely on weaker assumptions about the underlying distributions of the population. Such statistical procedures are referred to as *nonparametric* statistics. Chi-square, discussed in chapter 10, is one example of a nonparametric statistic. Nonparametric statistics often deal with variables that are less than interval in scale. At the heart of all statistical inferences, however, is the assumption of some sampling distribution against which observed outcomes are compared.

(*End of chapter* 8)

---

chapter

9

Hypothesis testing

At the conclusion of this chapter, you should be able to:

1. Name the two critical elements of a hypothesis: (a) a statement of an event (specific parameter value or specific parameter relationship) that is (b) set forth in such a way that the statement can be rejected.
2. Correctly label a set of statements as meeting or not meeting the definition of hypothesis.
3. Given an alternative or research hypothesis (hypothesis of interest), state an acceptable null hypothesis; given a null hypothesis, state two possible alternative hypotheses.
4. Given a null hypothesis, state whether a one-tailed or two-tailed test of significance is required.
5. Define Type I error in statistical decision making (to reject the null hypothesis when it is in fact true [false negative] [$p = \alpha$]).
6. Define a Type II error in statistical decision making (to fail to reject a null hypothesis when it is in fact false [false positive] [$p = \beta$]).
7. State the relationship between Type I and Type II errors (when the probability of a Type I error [α] gets smaller, the probability of a Type II error [β] gets bigger).
8. Given that the probability of wrongly rejecting the null hypothesis when it is true should be 6%, state that the level of significance, α, is equal to .06.
9. Define and discuss the relationships and differences among a *sample* distribution, a *sampling* distribution, and a *population* distribution.
10. Discuss the role played by each of the three types of distributions (sample, sampling, and population) in using inferential statistics.

1 There are five basic steps involved in doing all statistical tests. They are (1) making assumptions, (2) obtaining the sampling distribution, (3) selecting a significance level and critical region, (4) computing the test statistic, and (5) making a decision. Each of these steps requires some attention and can be used to introduce a number of concepts and operations that you may have seen previously used in other contexts.

There are certain assumptions that a researcher must make when approaching an experimental problem. Certain of these are assumptions about the *population* to which he wants to generalize and the *method of choosing the sample*. Some of these assumptions are those about which a researcher is relatively certain (he is willing to accept them) and other assumptions are those which he is dubious about accepting. It is the latter of these assumptions that he is interested in testing. From your knowledge of sampling procedures, it is possible to conclude that assumptions about the method of sampling are usually accepted (i.e., the researcher is certain about them), and they have little interest to the researcher when he proposes to:

A. Randomly select individuals from the population
B. Select only individuals from the population who he is sure will not have any influence on the outcome of the experiment

(A, *go to* 2)
(B, *go to* 3)

---

2 Using the method of random sampling would indeed enable the researcher to accept the assumptions about sampling procedures, which dictate that a sample should be random.

(*Go to* 4)

---

3 If individuals are selected on the basis of whether they will influence the results of an experiment, the procedure will produce a biased sample and probably biased results. Such a sampling procedure would show that the researcher was interested in the effects of the sampling procedure, and thus these assumptions would be in question.

Using the method of random sampling would enable the researcher to accept the assumptions about sampling procedures, which dictate that a sample should be random.

(*Go to* 4)

---

4 From the standpoint of performing a statistical test, it should be recognized that all assumptions have the same logical status. The statistical test can supply no information as to which of the assumptions is erroneous. Therefore the results of the experiment can be meaningful only if one of the assumptions is really in doubt. Usually the doubtful assumption is called the *hypothesis*. A statistical test is most meaningful if there:

A. Is only one hypothesis
B. Are several hypotheses

(A, *go to* 5)
(B, *go to* 6)

---

5 A statistical test is most meaningful only if there is one hypothesis, as you stated. This, of course, is because all assumptions have the same logical status. Each hypothesis represents a different assumption and, therefore, the fewer the hypotheses, the easier it is to pinpoint the faulty assumption because:

A. A statistical test does not tell which assumption is erroneous
B. It will be easier to calculate the statistical test

(A, *go to* 7)
(B, *go to* 8)

---

6 "Several hypotheses" is not the best answer. A statistical test is meaningful only if there is one hypothesis. All assumptions have the same logical status, and each hypothesis represents a different assumption. Therefore the fewer the hypotheses the easier it is to pinpoint the faulty assumption because:

A. A statistical test does not tell which assumption is erroneous
B. It will be easier to calculate the statistical test

(A, *go to* 7)
(B, *go to* 8)

---

7 A statistical test does not tell which assumption is erroneous. You are correct.

(*Go to* 9)

---

8 Ease of calculating a statistical test does not help you pinpoint the faulty assumption. The correct answer is that a statistical test does not tell which assumption is erroneous. Therefore the fewer the hypotheses the easier it is to pinpoint the faulty assumption.

(*Go to* 9)

---

9 The degree of confidence in accepting assumptions is usually provided by research methodology (i.e., research methodology involves, in part, the researcher's ability to randomly sample from a population) or a knowledge of previous research findings. These assumptions of relative certainty are usually lumped together and called the *model*. The doubtful assumptions concerning the population or method of sampling are called the:

A. Hypothesis
B. Statistic

(A, *go to* 10)
(B, *go to* 11)

---

10 The *hypothesis* is that assumption which is in doubt, and therefore it is the single assumption that the researcher will want to test.

(*Go to* 12)

---

11 The assumption in doubt is not a statistic. The *hypothesis* is that assumption which is in doubt, and therefore it is the single assumption that the researcher will want to test.

(*Go to* 12)

---

12 Even in the simple example of flipping a coin ten times, there were a number of assumptions:

1. The method of flipping constituted a random sample of all possible flips of a coin.
2. The events or flips were independent of each other.
3. The coin was an honest coin (it was not worn unevenly, etc.).

Normally the first two assumptions would be the model, and the hypothesis would be the third because usually our interest would center on whether the coin was honest. It should be pointed out, however, that the first assumption (the method of flipping constituted a random sample of all possible flips of a coin) could be doubtful because one could question the *method* of flipping the coin. Perhaps the person doing the flipping

Continued.

was not using a random procedure and was securing a bias by the way the coin was flipped. However, you were *interested* in whether the coin was biased and were therefore willing to *assume* that the other statements (assumptions) were not in error.

(*Go to* 13)

---

13 "The coin was an honest coin" is a statement of a hypothesis that the researcher would be interested in *rejecting*. This illustrates an important concept in the process of hypothesis testing. Theoretically, the researcher must always pose hypothesis statements that can be rejected as unacceptable. To be rejected, a hypothesis must be specific. The specific hypothesis to be tested is called the *null hypothesis* (symbolically represented by H_0). The usual null hypothesis is that the expected value of the statistic you are interested in is zero (no difference between groups, no relationship between variables, etc.). The null hypothesis, however, does not have to be a no-relationship statement as long as it is specific. Which hypothesis below is a correct null hypothesis?

A. Learning program X will lead to achievement gains.
B. Learning program X will have no affect on achievement gains.

(A, *go to* 14)
(B, *go to* 15)

---

14 Statement A would not be a good null hypothesis. "Learning program X will lead to achievement gains" suggests a general but unspecified relationship between achievement and learning program X. The null form should state a specific relationship, such as that no relationship exists.

(*Go to* 16)

---

15 Statement B would be a better null hypothesis. "Learning program X will have no affect on achievement gains" is a specific statement of *no* relationship between variables. Even this hypothesis, however, must be more specific to actually indicate the statistical test to be used.

(*Go to* 16)

---

16 The conservative nature of inferential statistics dictates that the researcher make it difficult for himself to reject the null hypothesis and thereby accept the research hypothesis.

When the hypothesis is in null form, a sampling distribution can be selected and evidence will thus be provided so that a decision can be made for either rejecting or not rejecting the null hypothesis. If the null hypothesis is rejected, it would be rejected in favor of the *research hypothesis*, which would be: "Learning program X will lead to achievement gains." The research hypothesis, which can be considered an alternative hypothesis to the null hypothesis, is usually the hypothesis in which we actually are interested. The researcher would hope the null hypothesis to be faulty so it could be rejected in favor of the alternative or research hypothesis. The alternative usually states a positive or negative relationship.

Which hypothesis below is a good example of a typical alternative or research hypothesis?

A. L-Dopa will have no effect on the patient's symptoms of Parkinson's disease.

B. Kline's syndrome is a genetically related condition.

(A, *go to* 17)
(B, *go to* 18)

---

17 You have selected an example of a typical null hypothesis. The null hypothesis usually states no difference between variables and is set forth in such a way that the statement can be rejected.

(*Go to* 18)

---

18 "Kline's syndrome is a genetically related condition" asserts a general relationship between the condition of Kline's syndrome and its cause, i.e., a genetic cause. It is best considered an alternative to a null hypothesis, which would state that Kline's syndrome is *not* a genetically related condition or that there is no relationship between Kline's syndrome and genetic factors.

(*Go to* 19)

---

19 Although in practice one usually starts with a research hypothesis and then states a null hypothesis, you should also be able to state a research hypothesis from a given null hypothesis. For instance, given the H_0: "Rational therapy will have no effect on anxiety as measured by the Taylor manifest anxiety scale," what would be a possible alternative or research hypothesis?

State a research hypothesis, then . . .

(*Go to* 20)

---

20 There could be several alternative or research hypotheses. You could have said:

H_1: Rational therapy will have an effect (either positive or negative) on anxiety.

H_2: Rational therapy will have a positive effect on anxiety.

H_3: Rational therapy will increase (have a negative effect on) anxiety.

Although several alternatives may be possible, the researcher usually constructs his research hypotheses on the basis of a theory. Therefore, given your knowledge of the effect of rational therapy on client behavior (i.e., given some theoretical expectation), you would probably have chosen H_1 or H_2 but not H_3.

The sampling distribution and the actual test statistic can only be computed when the hypothesis is stated in null form. In the coin example again, only when the coin is assumed to be honest can the binomial formula be used to calculate the exact probabilities. If one were to hypothesize that the coin was dishonest, it would be impossible to compute a sampling distribution because there are many ways in which the coin could be dishonest. Unless a specific value of dishonesty is determined, the distribution cannot be determined, and a researcher is often not in a position to do this. Therefore the null hypothesis would have to be that the coin was honest: i.e., H_0: In the long run there will be no difference between the number of heads flipped and the number of tails flipped.

(Go to 21)

21 The second step in performing a test of statistical significance and thereby testing a hypothesis is to obtain the sampling distribution. Such a distribution will tell how likely each of the possible outcomes of an experiment is *if the assumptions made are actually correct*. The values of the distribution in a coin-flipping experiment are best determined by:

A. Placing the possible outcomes on a normal curve

B. Applying the binomial formula

(A, *go to* 22)

(B, *go to* 23)

22 Placing the possible outcomes on a normal curve would not be the best method for determining the values of the sampling distribution for this experiment. In experiments in which the outcomes are just one of two possible states, it is usually best to apply the binomial formula. Sometimes the resulting distribution will appear symmetrical, like the coin flipping experiment, or it may even resemble a normal distribution. However, as we discussed earlier, there are many sampling distributions other than the normal distribution.

(Go to 24)

23 Since a coin flipping experiment involves an outcome with only two possible states (heads or tails), the values of the distribution are best computed using the binomial formula.

(*Go to* 24)

24 If the assumption is actually correct that the coin was honest in the coin flipping experiment, the likelihood of getting ten successive heads in ten flips of the coin would be 1 in 1024 attempts. This was calculated by using the binomial formula. In the hypothesis test, one is attempting to decide whether the experimental outcomes were so much different from the theoretically defined probability of the outcomes that the researcher should risk rejecting the null hypothesis (i.e., rejecting that the coin is honest). The word *risk* is used because:

A. The researcher could possibly get the 1 time in 1024 as a result of the nature of probability
B. The probability ratios are only rough estimates

(A, *go to* 25)
(B, *go to* 26)

25 You chose the correct answer. If the coin was flipped ten times and all heads came up, two things may have happened. The 1 chance in 1024 may have occurred during the first series of experiments (i.e., ten flips), or the assumption of the coin's honesty may be false. Because the likelihood of getting this result by chance is so small (1 in 1024), the researcher would normally *risk* rejecting the assumption. Usually experiments repeat trials because probability best explains events as they occur *in the long run*. However, a risk still exists no matter how many experiments are run.

(*Go to* 27)

26 Probability ratios are *exact* ratios (within limits of calculation accuracy) of the probability for occurrence. The correct answer was A. If the coin was flipped ten times and all heads came up, two things may have happened. The 1 chance in 1024 may have occurred during the first series of experiments (i.e., ten flips), or the assumption of the coin's honesty may be false. Because the likelihood of getting this result by chance is so small (1 in 1024), the researcher would normally *risk* rejecting the assumption. Experiments usually consist of repeated trials because probability best explains events as they occur *in the long run*. However, a risk still exists no matter how many experiments are run.

(*Go to* 27)

27 The reason for stating hypotheses in *null* form is that:

A. There is no theoretical interest in stating the research hypothesis
B. A probability distribution can then be used to describe the likelihood of events and thus provide a basis for decision making

(A, *go to* 28)
(B, *go to* 29)

---

28 You said that the reason for using the null form was that there is no theoretical interest in stating the alternative or research hypothesis. Actually, the research hypothesis is formulated on the basis of theoretical constructs (i.e., the expectations that the researcher always has prior to entering the experimental situation), and so it should be of strong theoretical interest.

The null form is used for the hypothesis test because it is a specific statement of a relationship. The statistic that describes the relationship can then be stated, and the sampling distribution for that statistic can be determined. The sampling distribution, described in terms of probability of occurrence for any event, provides the basis for decision making—the decision of whether to reject the null hypothesis.

(*Go to* 30)

---

29 The null form is used for the hypothesis test because it is a specific statement of relationship. Consequently, as you stated, a probability distribution can then be used to describe the likelihood of events and thus provide a basis for decision making. The decision to be made is whether to reject the null hypothesis.

(*Go to* 30)

---

30 The third step in statistical tests, once the assumptions have been identified and a sampling distribution has been obtained for the null hypothesis (the actual selection of a sampling distribution will be left to a later time), is to select a *significance level* and determine the *critical region*. This involves making a decision on a set of alternatives which, should they occur, would require a rejection of the null hypothesis. The *set* of alternatives or outcomes that would warrant the rejection is called the critical region. These outcomes would warrant rejection of the null hypothesis because:

A. The probability of these outcomes occurring by chance is low
B. The results of the experiment have shown the hypothesis to be false

(A, *go to* 32)
(B, *go to* 31)

---

31 Your answer was B, "the results of the experiment have shown the hypothesis to be false." Generally the critical region is determined before the experiment is performed and results made available. The decision about what alternatives comprise the critical region should be made prior to the actual experiment. The set of alternatives in the critical region, therefore, would not be taken from the results of the experiment.

(*Go to* 32)

32 Rejection of the null hypothesis is warranted when the probability of the outcomes occurring by chance is low. The critical region becomes one of the two categories of possible outcomes: (1) those outcomes which lead the researcher to reject the null hypothesis (the critical region), and which by definition have a small chance of occurring; and (2) those outcomes whose occurrence would not permit rejecting the null hypothesis.

(*Go to* 33)

33 Fig. 9-1 is the probability distribution for getting heads in ten flips of a coin. What alternatives (outcomes) would best be considered the critical region in Fig. 9-1?

A. The outcomes in region B
B. The outcomes in region A *or* C
C. The outcomes in regions A *and* C

(A, *go to* 34)
(B, *go to* 35)
(C, *go to* 36)

FIG. 9-1

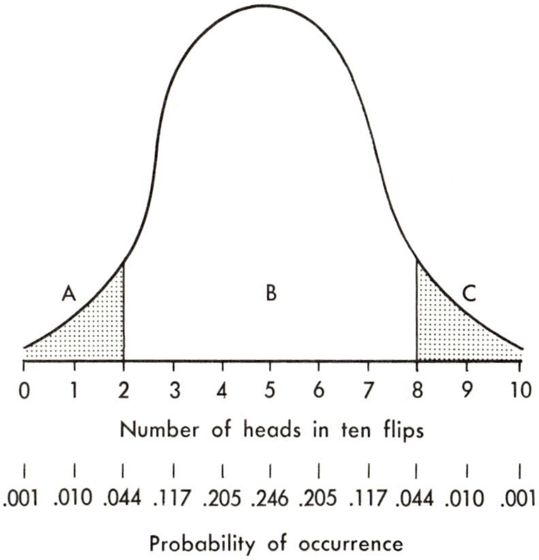

34 Actually, you selected the probability ratios or outcomes that have a relatively high likelihood of occurring. The critical region is comprised of those outcomes with the lowest probability of occurrence.

Try again.

(*Go to* 33)

---

35 Technically you are correct. You selected the region A *or* C as having a set of alternatives with low probability of occurrence. However, both regions A and C have outcomes with low probability of occurrence. Since the critical region is concerned with those values of the statistic (in this case number of heads) that occur only rarely, unless other assumptions are made, both regions A *and* C should be in the critical region.

(*Go to* 37)

---

36 Your choice, which was correct, was the set of probabilities in areas A *and* C of the probability distribution. Notice that ratios at both ends of the distribution have a low chance of occurring. If other assumptions were made, it would also be possible to select only the alternatives on one end of the distribution as being in the critical region.

(*Go to* 37)

---

37 Logically there must be a basis for choosing how many probability ratios should be included in the critical region. One does not select the cutoff point for the critical region strictly on an arbitrary division of the distribution. There are two considerations that the researcher must make when deciding how large or small to make the critical region:

1. Whether both ends (or tails) of the distribution should be included in the critical region
2. How the critical region of a certain size will contribute to the chances for (a) rejecting the null hypothesis when it is in fact true or for (b) failing to reject the null hypothesis when it is actually false

(*Go to* 38)

---

38 Take the second consideration first: the errors of rejecting a true null hypothesis or failing to reject a false null hypothesis. These are frequently referred to as making what is called a Type I or Type II error.

Errors can occur because we are moving from a general statement about a population, expressed in our hypothesis, to a specific instance in our sample. The hypothesis is a statement of what we expect. The logic behind our inference is based on the mathematics of probability for what happens in the long run. In any conclusion based on probabilistic reasoning, there is a chance that the conclusion is wrong.

The logic of our null hypothesis rejection is that the null hypothesis states an expected value or outcome in our population. The sampling distribution, however, indicates that even if the null hypothesis is true, a variety of outcomes could occur in our sample by chance (by variation from sample to sample). In the sampling distribution, those events least likely to occur are those plotted in the tails of the curve. At some point in the distribution of possible outcomes the researcher says, "These events and those events even less likely to occur are so rare a set of events that if one of them occurs in my sample, I will have to conclude that my null hypothesis was wrong." The sum of all the probabilities for all the events that lead to rejection of the null hypothesis is the *significance level* or *alpha* (α) *level*. Note, however, that even if one of these rare events does occur, it could still be true, although it is unlikely, that the null hypothesis is correct. If the null hypothesis is indeed correct and we reject it, then we have made a Type I (or alpha) error.

Thus the possibility for a Type I error:

A. Will always remain no matter how large or small the critical region is

B. Will exist only when the critical region is large (i.e., a large portion or many alternatives at either end of the probability distribution are included in the critical region)

(A, *go to* 39)
(B, *go to* 40)

---

39 The possibility of committing a Type I error will always exist as you indicated, as long as conclusions are reached on the basis of probability theory. The chances for having a Type I error, however, can be controlled by changing the size of the critical region. The *smaller* the critical region, the *less* chance there is for a Type I error.

(*Go to* 41)

---

40 Actually, there will always be a possibility of a Type I error as long as conclusions are reached on the basis of probability theory. The answer

Continued.

you selected does intimate that a relation exists between a Type I error and the size of the critical region. This is true because the smaller the critical region, the less chance (or probability) there is of committing a Type I error.

(*Go to* 41)

---

41 Mathematically, the chance (or probability) of making a Type I error in an experiment is equal to the sum of each of the probability ratios in the critical region. In our coin flipping experiment, if the critical regions consist of zero and one head at one end of the distribution of probabilities and nine and ten heads at the other end, then the probability of a Type I error would be:

$$\frac{1}{1024} + \frac{10}{1024} + \frac{10}{1024} + \frac{1}{1024} = \frac{22}{1024} \text{ or about } .022$$

The probability for a Type I error would be decreased the *most* (i.e., would be minimal) in the coin flipping experiment if the critical region contained (refer to Fig. 9-1):

A. 0 and 10 heads as outcomes in ten flips
B. 1 and 9 heads as outcomes in the ten flips
C. 0, 1, 9, and 10 heads

(A, *go to* 42)
(B, *go to* 43)
(C, *go to* 44)

---

42 If only a 0 and 10 were included as outcomes in the critical region, the size of the critical region would be minimal, and therefore the probability of committing a Type I error would be small.

(*Go to* 45)

---

43 You said that the probability of committing a Type I error would be smallest if only the outcomes 1 and 9 were included in the critical region. That is not correct. The complete end or "tail" of the probability distribution is included in the critical region so that the 0 and 10 must also be in the region. Furthermore, if only the outcomes 1 and 9 heads were included, it would not represent the critical region with the lowest probability for a Type I error. The probability ratio for 1 head is .010; the same ratio obtains for 9 heads. Therefore the probability for a Type I error would be .010 + .010 = .02; the probability for 0 and 10 outcomes (choice A) would be .001 + .001 = .002. Thus you can see that the probability for a Type I error is minimized by selecting only 0 and 10 heads as outcomes in ten flips.

(*Go to* 45)

---

44 The answer you chose would be the same critical region used in the previous example with a significance level of .022. The critical region can be made smaller. The correct answer was A, only 0 and 10 heads as outcomes in the critical region. The probability for a Type I error would then be .001 + .001 = .002.

(*Go to* 45)

---

45 The probability of committing a Type I error is referred to as the *significance level* of the hypothesis test. A researcher can have a significance level that would minimize the chances of a Type I error by choosing a critical region with low probability ratios. However, there is then the additional risk of making yet another statistical error, a *Type II error*. The researcher may fail to reject a false hypothesis, especially as he/she decreases the chances for a Type I error. If a set of assumptions (hypotheses) is indeed false, there is less and less likelihood of rejecting them as the researcher makes the significance level lower. Therefore as the researcher endeavors to decrease the chances for a Type I error:

A. The possibility of committing a Type II error increases
B. Type II errors will also be decreased

(A, *go to* 46)
(B, *go to* 47)

---

46 By lowering the significance level, the researcher minimizes the risk of a Type I error and consequently increases the risk of committing a Type II error. You are correct. There is an inverse relationship between the probability of committing the two types of errors, and it is, therefore, impossible to minimize the risks of both errors simultaneously.

(*Go to* 48)

---

47 The probability of committing a Type II error will not be decreased. By lowering the significance level (thereby minimizing the risk of committing a Type I error), the researcher will *increase* the risk of a Type II error. There is an inverse relationship between the probability of committing the two types of errors, and it is, therefore, impossible to minimize the risks of both errors simultaneously.

(*Go to* 48)

---

48 A Type I error can be made by:

A. Rejecting a hypothesis when the hypothesis is indeed true

B. Failing to reject a hypothesis when the hypothesis is indeed false

(A, *go to* 49)
(B, *go to* 50)

---

49 You have selected the correct definition of a Type I error; i.e., rejecting a hypothesis when in fact it is true. It is easy to confuse the two kinds of errors.

You should be able to tell when a significance level will increase the probability for either type of error. Which of the following choices will decrease the chances of committing a Type II error?

A. Changing the significance level from .002 to .02

B. Selecting a significance level of .01 instead of .20

(A, *go to* 55)
(B, *go to* 54)

---

50 You have selected the definition for a Type II error. A Type I error is made when the researcher rejects a hypothesis that is in fact true. It is easy to confuse the two errors.

In which example below is there the greatest probability of committing a Type II error?

A. A researcher concluded that his/her null hypothesis could be rejected in favor of the research hypothesis because the results occurred within the critical region of .56.

B. A researcher said his/her experiment was a failure because the research hypothesis could not be substantiated by rejecting the null hypothesis at a significance level of .001.

(A, *go to* 51)
(B, *go to* 52)

---

51 By setting an alpha level at .56, almost any outcome is going to lead the researcher to reject the null hypothesis even if it is true. You selected the example that maximizes the risk of committing a Type I error, not a Type II error. The Type II error is made when the researcher fails to reject the null hypothesis when indeed the null hypothesis is false. Therefore a Type II error has the greatest probability of being committed by the researcher who says the experiment failed because the research hypothesis could not be substantiated (by rejecting the null hypothesis) at a significance level of .001. The significance level (Type I error) is low and the probability of committing a Type II error is higher if low significance levels are used.

(*Go to* 53)

---

52 You correctly selected the example that maximizes the risk of committing a Type II error. When the significance level is small it is hard to reject the null hypothesis, and even if the null hypothesis is false you are likely to fail to reject it. Therefore with a small critical region (e.g., .001), a Type II error might be suspected.

(*Go to* 53)

---

53 Which of the following choices would decrease the probability of committing a Type II error?

A. Change the significance level from .002 to .02
B. Select a significance level of .01 instead of .20

(A, *go to* 55)
(B, *go to* 54)

---

54 If the significance level were changed from .20 to .01, there would be an increase in the probability of committing a Type II error. Since fewer outcomes or results could occur in the critical region, the chances of obtaining evidence for rejecting the null hypothesis are decreased, and the researcher will have a more difficult time in rejecting a hypothesis that was indeed false.

(*Go to* 56)

---

55 The chances for committing a Type II error are indeed decreased if the significance level were to be enlarged from .002 to .02. By selecting a higher significance level, the chances for failing to reject the null hypothesis when it is actually false are not as great because more outcomes or results could fall within the critical region.

(*Go to* 56)

---

56 The following paradigm should help you remember the two types of statistical error that can be committed in inferential statistics.

		Decision to reject or not reject the tested (null) hypothesis	
		Not reject	*Reject*
Actual situation regarding tested (null) hypothesis	*True*		Type I error = α
	False	Type II error = β	

If the probability of committing a Type I error were calculated to be .06, what would be the significance level for this hypothesis test?

A. .05

B. .06

(A, *go to* 57)

(B, *go to* 58)

---

57 The probability of committing a Type I error also indicates the significance level. Therefore if .06 is the probability of committing a Type I error, then the level of significance is also .06. The significance level is usually written with the symbol alpha (α). Therefore in this example, $\alpha = .06$.

(*Go to* 59)

---

58 Yes, .06 is the significance level for a test of a hypothesis and is the same as the probability of committing a Type I error. Usually the significance level is symbolized as alpha (α). Therefore in this example, $\alpha = .06$.

(*Go to* 59)

---

59 Unfortunately, it is difficult to actually compute the probability of committing a Type II error. Most researchers will set their probability for a Type I error at a certain level and then try to select a statistical test that minimizes the risk of a Type II error. The ability of a statistical test to eliminate false hypothesis is referred to as the *power* of a test. The power of a test, related to the probability of a Type II error, depends on which alternative hypothesis is actually correct. The problem of determining the probability of a Type II error is beyond the scope of this text, but the reader of statistical literature should be aware that not all tests of statistical significance are equally powerful. That is, although we can get any test to produce Type I error (rejecting a true hypothesis) α percent of the time, each test will vary in its power to reject a false hypothesis.

The researcher becomes a self-imposed taskmaster when selecting the critical region. To avoid biasing his/her results or favoring expected results, the researcher leans over backward to prove himself/herself wrong or to obtain results he/she actually does not want to obtain. The conservative nature of inferential statistics can be seen in this remark. In most cases a null hypothesis is set up with the hope of rejecting it in favor of the research hypothesis. Since in reality the researcher would like to reject the null hypothesis in favor of the research hypothesis, he/she should make it difficult to achieve this result by:

A. Choosing a large critical region
B. Choosing a small critical region
C. Guarding against a Type II error

(A, *go to* 60)
(B, *go to* 61)
(C, *go to* 62)

---

60 You selected A, choosing a large critical region. This is wrong because a large critical region would increase the number of likely results that could occur in the critical region. By so doing, the researcher would have a greater probability of rejecting the null hypothesis than had a smaller critical region been chosen. The correct answer would be to choose a small critical region, thus making it difficult to reject the null hypothesis, or the hypothesis that the researcher really does not want to end up with.

Most researchers use significance levels of .05, .01, or .001 because they represent rather small critical regions. The fact that it is .05 instead of .04 or .01 instead of .02 is of little concern, and specific values are arbitrary. The important point is that all represent small critical regions. (See Skipper et al., 1967.)

(*Go to* 63)

---

61 Your answer, B, is correct. By choosing a small critical region, the researcher makes it difficult to reject the null hypothesis, or the hypothesis that he/she really does not want to end up with. It is for this reason that significance levels of .05, .01, and .001 are often used in research literature. The fact that .05 is used instead of .04 or .01 instead of .02 is the result of a somewhat arbitrary decision. The important point is that all represent small critical regions. (See Skipper et al., 1967.)

(*Go to* 63)

---

62 Your choice, C, is incorrect. If one would guard against a Type II error, the critical region should be increased. A Type II error is a *failure* to reject the null hypothesis when, in fact, the null hypothesis is *false*. One measure against making a Type II error is to increase the size of the critical region, thus making it more likely that the null hypothesis will be rejected. But this also increases the chances of committing a Type I error. It is better to select a statistical test that is more appropriate for the situation. The best answer to the previous question would have been B; choose a small critical region, since this makes it difficult to reject the null hypothesis, or the hypothesis that the researcher really does not want to end up with.

It is for this reason that most researchers use significance levels of .05, .01, or .001; they represent small critical regions. The fact that .05 is used instead of .04 or .01 instead of .02 is of little concern. The specific values are arbitrary. The important point is that all represent small critical regions. (See Skipper et al., 1967.)

(*Go to* 63)

63 A second consideration in deciding what size to make the critical region is whether both ends (or tails) of the sampling distribution should be included. This decision is based on whether we are predicting the *direction* in which the results will occur. Many times the researcher is interested only in deviations that should occur in certain directions. In the coin flipping experiment, if the prediction were made that the coin was dishonest in favor of heads, the end of the distribution relating the probability of getting excessive numbers of heads would be adequate for deciding the coin's honesty.

As another example, it may be predicted that a vitamin A additive will produce faster healing for skin infections than a vitamin D additive. The prediction that A is faster than D, based on some theoretical reasoning, indicates the direction expected in our outcomes. The result is that the critical region for the hypothesis test will be at only one end of the probability distribution. When such theoretical reasoning about direction occurs, it is said that the statistical test is a *one-tailed* test.

Whenever direction has been predicted, one-tailed tests are preferable to two-tailed tests. However, the significance level (probability of committing a Type I error) will be the same for both one-tailed and two-tailed tests. From the distribution in Fig. 9-2, it can be seen that when a one-tailed test is used, the critical region is concentrated at one end of the distribution and forms a larger region for that tail. The significance level (i.e., the probability of committing a Type I error) in this case would be:

A. p = .025

B. p = .05

(A, *go to* 64)
(B, *go to* 65)

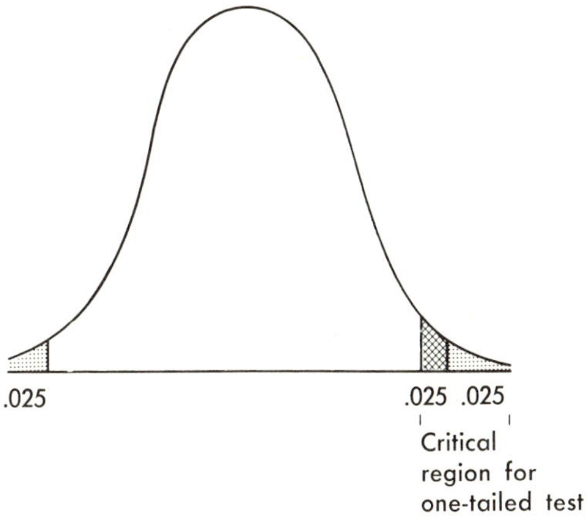

FIG. 9-2

---

64 Your answer, p = .025, is not correct. Fig. 9-2 shows that two areas, each with probabilities of .025, are included in the critical region. Thus the significance level is p = .05. When a one-tailed test is made, usually the researcher includes a larger area at one end of the sampling distribution (rather than smaller areas at each end) and keeps alpha at .05 or at whatever level has been selected.

(*Go to* 66)

---

65 You are correct; .05 is the significance level or critical region. Thus the probability of committing a Type I error usually is left constant regardless of whether one or two tails are used.

(*Go to* 66)

---

66 Which of the following null hypotheses would require a one-tailed test?

A. There is no difference in noise sensitivity between persons raised in inner cities and persons raised in suburban fringes.

B. Primary grade students in DISTAR reading programs will score on reading tests as well as or lower than students in conventional reading programs.

(A, *go to* 67)
(B, *go to* 68)

67 You said that a one-tailed test would be needed to test if there were any difference in noise sensitivity between two samples. That is wrong. Differences could occur in either direction; that is, inner city persons could be more sensitive or less sensitive than suburban persons. As Fig. 9-3 indicates, both tails would indicate a difference, and we are interested in any difference. Therefore a two-tailed test would be needed.

(*Go to* 69)

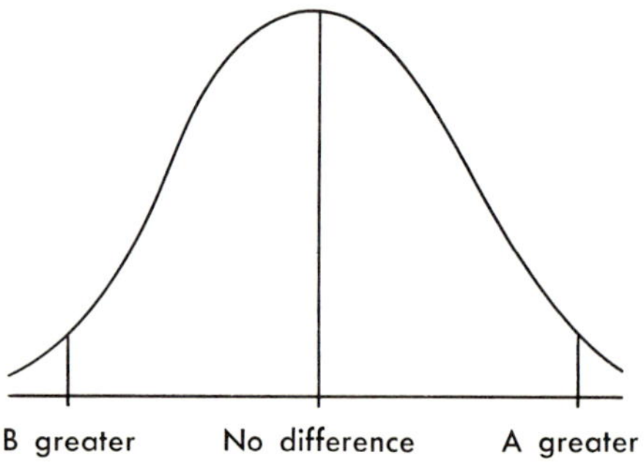

Fig. 9-3. Differences between groups A and B on any variable.

68 Hypothesis B stated there would be no difference or a negative difference between DISTAR readers and conventional readers. As Fig. 9-3 indicates, only a difference at one end of the distribution would allow us to reject the hypothesis. Thus we are interested in the direction of difference and would need a one-tailed test.

(*Go to* 69)

69 Given the following research hypothesis, would you use a one-tailed or a two-tailed test?

Air pollutants will cause detrimental irregularities in the functioning of the spleen by inhibiting the breakdown of red blood corpuscles that set free hemoglobin.

A. One-tailed test
B. Two-tailed test

(A, *go to* 70)
(B, *go to* 71)

---

70 A one-tailed test should be used. The fact that inhibition of corpuscle breakdown is identified illustrates that the direction in which irregularities will occur has been predicted.

(*Go to* 72)

---

71 A two-tailed test should not be used. If the research hypothesis stated only that air pollutants will cause irregularities in the function of the spleen, there would be no predicted direction for the deviation (they could be positive, neutral, or damaging irregularities), and a two-tailed test would be used. However, the hypothesis states that inhibition of corpuscle breakdown will occur, thus illustrating that the direction of the deviation (detrimental irregularities) has been predicted. This situation, therefore, dictates the use of a one-tailed test.

(*Go to* 72)

---

72 Three of the five steps used in carrying out a hypothesis test have been discussed. They are (1) making assumptions (model and hypothesis), (2) obtaining a sampling distribution, and (3) selecting a significance level and critical region. There remain yet (4) the computation of the test statistic and (5) a decision of whether to reject the null hypothesis.

The fourth step of actually computing test statistics will be covered more thoroughly in a later chapter. At this point it is only important to recognize what the test statistic is, what it does, and how it differs from the kind of statistics presented earlier as descriptive statistics.

(*Go to* 73)

---

73 A test statistic is a statistic that does not describe or represent a summary of some characteristic in a sample or population. In other words, a test statistic is *not* a mean, standard deviation, or any form of descriptive summarization of data. Rather it is a computation of a quantity from sample data that can be compared with a set of results that might

Continued.

be predicted by the sampling distribution. In comparing the results of the computation (i.e., the test statistic) with what should have happened according to probability theory (sampling distribution), a decision about the null hypothesis can be made. Examples of test statistics are the chi-square test (χ^2), t tests, and Z tests.

The test statistic itself is a rather simple mathematical procedure. In the coin flipping example, all one really needs to know is the probability distribution (i.e., sampling distribution) for the number of successes in N trials for a set of coin flips and the outcome of one experiment (i.e., flipping the coin ten times). A comparison can then be made with the sampling distribution to see if the results of the experiment fall within the critical region. If the results fall within the critical region:

A. The null hypothesis can be rejected
B. The research hypothesis is concluded to be true
C. The researcher will fail to reject the null hypothesis

(A, *go to* 74)
(B, *go to* 75)
(C, *go to* 76)

---

74 You selected A—the null hypothesis can be rejected. That is correct. The result, according to the test statistic, indicates that the probability for the outcome is so low that, by chance alone, it is improbable that this outcome would happen if the null hypothesis were true.

(*Go to* 77)

---

75 You selected B—the research hypothesis is concluded to be true. That is not correct. One can never say that the research hypothesis is true because the rejection of the null hypothesis might have been the result of causes other than the causes postulated in your research hypothesis. Therefore it *is* appropriate to say that the null hypothesis can be rejected *in favor* of the research hypothesis, and that the research hypothesis will be allowed to stand until someone can do a more conclusive test to disprove this hypothesis. The only conclusive statement is that the null hypothesis can be rejected (since the results produce a test statistic that has a probability of occurrence so low that it most likely would not happen by chance alone).

(*Go to* 77)

---

76 You selected C—the researcher will fail to reject the null hypothesis. This is an incorrect conclusion. The critical region is that region of the probability distribution which illustrates those outcomes whose probability of occurrence is so low that they would not be likely to occur by chance alone. The test statistic in the example shows results as belonging in the critical region. This means the events observed were rare events, if the null hypothesis were true. Rather than assume that rare events happened, we reject the null hypothesis in favor of some alternate hypothesis that can explain the observed events.

(*Go to* 77)

---

77 The test statistic is:

A. Arrived at by computing data from the population
B. Computed by applying a mathematical formula, such as chi-square, to the sampling distribution
C. Calculated as a descriptive summary of sample data
D. Computed from values obtained in a sample to be compared with the sampling distribution

(A, *go to* 78) (C, *go to* 80)
(B, *go to* 79) (D, *go to* 81)

---

78 You said that the test statistic is arrived at by computing data from the population. That is incorrect. The *population* is not the source of data per se. It is the *sample* that provides quantities used in the computation. It is the very role of inferential statistics to *infer* from a sample to the population. This is, of course, the characteristic of any statistic—it is derived from a sample. When descriptive values are derived from an entire population, it is called a parameter.

(*Go to* 81)

---

79 No, the test statistic is computed from quantities obtained in a sample, not from a sampling distribution. The sampling distribution will have been determined before the test statistic is obtained so that once the sample data are collected and the computation is made, the resulting test statistic can be compared with the values on the sampling distribution to see whether the test statistic will fall in the critical region.

(*Go to* 82)

80 The test statistic is *not* a descriptive summary of sample data. It is a mathematical value that relates the results of quantities from sample data to the sampling distribution for sets of outcomes possible in the experiment. It is a statistic that is of no inherent interest descriptively, but which is used in testing hypotheses.

(*Go to* 81)

81 The test statistic is computed from values obtained in a sample and relates the observed outcome to the expected outcomes of the sampling distribution. The test statistic allows you to decide whether the results fall within the critical region of the sampling distribution for that experiment.

(*Go to* 82)

82 Once the decision to reject or not to reject the null hypothesis has been made on the basis of the test statistic, that is, on whether it falls within the critical region of the sampling distribution for that experiment, the fifth and last step in the hypothesis test is completed. If the null hypothesis is rejected in favor of the research hypothesis, the results are usually reported to be significant at the predetermined level declared as the significance level. Thus the report may say, "Treatment A group was significantly better than treatment B group at the .05 level." This wording is often used even though the significance level actually applies to rejecting the null hypothesis, rather than confirming the research hypothesis.

(*Go to* 83)

83 In inferential statistics we are usually interested in comparing the results of some sample data with what we believe the population to be like as expressed in our hypothesis. We may be interested in indicating the likely range of values within which a population parameter would fall, or we may be interested in a specific value that we hypothesize to exist. In either event we cannot infer directly from the sample to a population with a known probability of accuracy. Instead we ask, "If the population were as we expect (hypothesize), what would be the probability of finding the outcome observed in our sample?"

There is a *population distribution* for every variable or set of variables in the population. The distribution is usually unknown because we cannot actually measure all of most populations. Our hypothesis, however, concerns only one statistic about one distribution. It may be the mean, or the difference between the means of two populations, or the correlation between two variables in the population. There also is a *sample distribution* for every variable or set of variables in the sample. Again, however, we are interested in only one statistic from the sample.

(*Go to* 84)

---

84 When the statistic in which we are interested is specified in the hypothesis, then we can specify the distribution of that statistic—the *sampling distribution.* If the researcher should assume the wrong sampling distribution, invalid conclusions will follow. For the most part the applied researcher works with statistics for which the distributions are well known, the most common being the normal distribution. The sampling distribution, you will recall, represents the probability distribution of all possible outcomes of a statistic, given the hypothesis and specific sample. As Fig. 9-4 indicates, using the mean as an example, our sampling distribution lets us compare our observed $\overline{X}$ with all the possible means given a hypothesized μ. The test statistic is the mathematical means by which we make our comparison. If our observed value ($\overline{X}$) is too rare an event, then we reject our hypothesized value (μ) and look for an alternative hypothesis.

(*End of chapter 9*)

FIG. 9-4

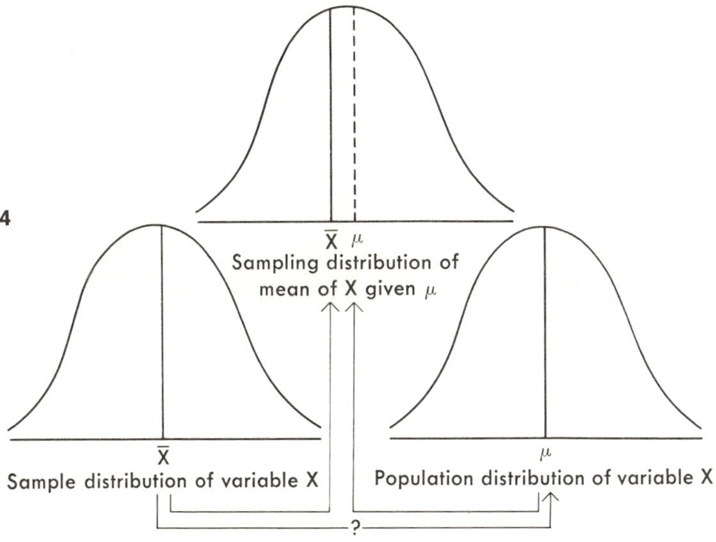

chapter
10

Tests of statistical significance

At the conclusion of this chapter, you should be able to:

1. State which test of statistical significance is appropriate, depending on the data available, scale of measurement, sample size, and the hypothesis to be tested, when given problems involving one sample and one variable, where the basic question is: Are the values observed those which were expected?
2. State which test of statistical significance is appropriate, depending on the data available, scale of measurement, sample size, and the hypothesis to be tested, when given problems involving two (or more) samples and one variable, where the basic question is: Are the samples alike?
3. State that the basic inferential tests are all forms of the ratio:

$$\frac{\text{Obtained statistic} - \text{Expected statistic}}{\text{Standard error of statistic}}$$

   This ratio can be stated as the distance between observed and expected values expressed in standard error units.
4. Select an appropriate test of statistical significance according to the information available in the problem (mean vs. proportion, standard error of population vs. estimated standard error and size of N), when given a null hypothesis that a sample mean or proportion is equal to some expected population value.
5. Select an appropriate test of statistical significance according to the information available in the problem (mean vs. proportion; small samples vs. large samples), when given a null hypothesis about differences between two samples on some variable.
6. Select from a list of assumptions those that apply to a given Z or t test.
7. Compute a Z or t value and the degrees of freedom associated with the problem when given a Z or t formula and the appropriate data.

8. Compare the t and normal distributions with respect to shape. (The normal distribution is the same for any size N, whereas the t distribution is *different* for each N; the t distribution is *flatter* than the normal curve for small N and *approaches the normal curve* as N gets large.)
9. Evaluate any Z or t test calculated in objective 7 by rejecting or failing to reject the associated null hypothesis at a stated α level, when given a normal table and a t table.
10. Lay out the appropriate contingency table or state that a contingency table is not appropriate for a given hypothesis.
11. Discriminate between contingency problems in which the data are obtained from dependent and independent samples, and state which are appropriate for a chi-square analysis.
12. Determine the expected frequencies for each cell in a contingency table when given a set of two-variable frequency data as a contingency table and a null hypothesis of independence of the variables.
13. Calculate the chi-square value for the contingency table produced in objective 12.
14. Determine the degrees of freedom associated with the chi-square values calculated for item 13.
15. State the significance of a specified chi-square value given a chi-square table, α, and the appropriate degrees of freedom.
16. State (without calculating) which table will produce the larger chi-square value when given two contingency tables that are identical except that the frequencies in one are n times larger than the frequencies in the other.
17. When given a null hypothesis, an alternative hypothesis, and a requirement for a 95% confidence level:
    a. State the required significance level α
    b. Determine whether to reject the null hypothesis given that the observed values could have occurred X% of the time by chance
18. State the general form of an equation for calculating a confidence interval and explain the equation using the concepts "statistic," "sampling distribution," and "standard error."
19. Calculate the 95% confidence interval for a population mean given $\overline{X}$, s, N, and t.
20. State an interpretation of the confidence interval in meaningful English sentences when given a case with a population estimate based on a sample and a confidence interval for that estimate.

---

1 In previous chapters you were introduced to the logic of tests of statistical hypotheses. However, little was said about the ways one can compute the test statistic.

You should recall that the basis of the test statistic is that it represents a comparison of a quantity computed from sample data to a set of results that might be predicted by the sampling distribution. Thus by comparing

Continued.

these results with what should have happened according to probability theory (i.e., the sampling distribution), one can decide whether to reject the null hypothesis.

A test statistic is really nothing more than some form of a comparison of a sample characteristic (e.g., $\overline{X}$) with some form of the population parameter that the sample statistic is to represent (e.g., μ). Thus a test statistic would be exemplified by a comparison of:

A. The mean score made by all sociology majors on the Graduate Record Examination with the mean score made by a randomly selected group of graduates from a state university

B. The mean Graduate Record Examination score made by a random sample of all sociology majors with the mean score of all sociology majors taking the examination

(A, *go to* 2)
(B, *go to* 3)

---

2 A test statistic would not show a comparison between all sociology majors on the examination with a randomly selected group of graduates. Your answer is incorrect. The sample of graduates need not be from the population that is defined by the major, sociology. In other words, the sample that is used for the comparison must be from the population with which it is being compared.

(*Go to* 4)

---

3 The mean examination score of all sociology majors compared with the mean score from a sample of that population parameter will exemplify the function of a test statistic. The sample that is used for the comparison must be from the population with which it is being compared.

(*Go to* 4)

---

4 Since the intent of tests of statistical significance is to compare the similarity of the statistic and the parameter, some measure of the *amount of dispersion* found in the distribution of all the possible sample statistics must be incorporated into this comparison. The statistical vehicle for measuring dispersion of all possible sample statistics is:

A. The standard deviation of the sampling distribution

B. The standard deviation of the sample

(A, *go to* 5)
(B, *go to* 6)

---

5 The correct answer is the standard deviation of the sampling distribution (or the standard error of the statistic). The comparison of the statistic and the parameter is stated in terms that are relative to the dispersion that is found in the sampling distribution of the particular statistic of interest. A comparison of the statistic and the parameter relative to the distribution of the sample characteristics is called the *test statistic*. You will see the test statistic in various forms in this chapter. However, they will all have essentially the same characteristics described by the above definition.

(*Go to* 7)

---

6 The standard deviation of the sample will not allow for an adequate comparison of the statistic with the parameter. The standard deviation of the sample(s) is only a statement of the amount of spread in the sample.

(*Go to* 5)

---

7 All of the parametric tests of statistical significance discussed in this chapter are ratios of the general form:

$$\frac{\text{Obtained statistic} - \text{Expected statistic}}{\text{Standard error of statistic}}$$

Recall that the standard error of a statistic is the standard deviation of the sampling distribution for that statistic. Thus the above ratio indicates:

A. How closely the obtained statistic relates to the expected statistic
B. That to use the Z and t tests, the population statistic or parameter must be known

(A, *go to* 8)
(B, *go to* 9)

---

8 The ratio will indicate how many standard-error-of-the-statistic units the obtained statistic is away from the expected statistic. Your answer is correct. The population statistic may not always be known when using this ratio. Later, you will be shown how Z and t tests are used when the population statistic is not known.

(*Go to* 10)

9 It might appear by logical deduction from the equation given in frame 7 that the expected statistic could only be derived from the population parameter. However, it is not true that using the ratio always would require direct knowledge of the parameter. Later you will be shown how Z and t tests are used when the population parameter is not known.

The correct answer is that the ratio indicates the number of standard-error-of-the-statistic units the obtained statistic is away from the expected statistic.

(*Go to* 10)

10 Each statistic that is chosen has associated with it a specific sampling distribution, i.e., a distribution of the probabilities of the occurrences of each of the possible samples. Therefore the choice of which statistic to use must be made in terms of the desired shape or nature of the sampling distribution. The shape or nature of the sampling distribution can be ascertained (at least theoretically) by noting the nature of the statistical problem at hand. All of the Z and t tests in this chapter deal with means, proportions, and differences between means and proportions. These statistics all have normal or approximately normal sampling distributions.

When choosing a test statistic, the following information about the experiment or problem at hand will help to identify the appropriate test:

1. Size of the sample (N)
2. Nature of the descriptive statistic (usually an $\overline{X}$ or proportion)
3. Whether the population parameters are known or must be estimated
4. Whether interest is in comparing two sample statistics or comparing a single sample statistic to a population parameter

Each of the points will be discussed with each of the statistics presented so you will have an opportunity to review them again. Much of your knowledge regarding the appropriate use of statistics will be based on these distinctions.

Before proceeding into this chapter, you should be apprised of two problems.

1. The use of tests of statistical significance for the testing of hypotheses presents the researcher with a series of dilemmas. The most important dilemma is that the researcher must make several decisions at each step of the testing process. One of the most critical of these decisions is that of using the correct test in the correct situation. To aid you in this process, this chapter is organized so that the major assumptions that are required for each test are stated. You should acquire a clear understanding of them.

2. There is always a danger in confusing statistical significance with substantive significance. Because a relationship or test statistic proves

to be statistically significant, this does not mean that it will be substantively significant. We know, for example, that there is a statistically significant relationship between the number of letters in the name of a month and the mean monthly temperature of that month. (As the temperature increases, the number of letters in the month's name decreases.) Although this finding is of statistical significance, it is of little substantive significance (Sargent, 1955). *(Go to 11)*

---

11 Two of the most often used test statistics are the Z test and the t test. Fundamentally, the decision for using the Z test or t test is based on the size of N; i.e., the size of the sample that the researcher has. When samples are *greater* than 30, the Z test is used. When N is *equal to* or *less than* 30, the t test is used. (You should recall what the central limit theorem indicates about sampling distributions based on the size of the sample.)

Because it is a bit simpler to understand, and because it follows more directly from knowledge you already have, the Z test will be presented first. The Z test is really nothing more than a conversion of the descriptive statistic for the sample into a standard score. Through this conversion procedure it becomes possible to use the *normal table* as a representation of the sampling distribution for the population. This procedure is not unlike finding Z scores for areas under the normal curve.

The normal table is a listing of the probabilities of the occurrence of all of the possible sample outcomes (converted into standard [normal] scores) and is in the shape of a normal distribution. Through its use it is possible to determine whether the sample statistic that the researcher obtains through empirical research is close to or distant from the hypothesized population parameter that it is supposed to represent. How distant the Z statistic must be to reject the H_0 depends on the significance level that was already determined by the researcher.

Four types of Z tests are discussed in this chapter. If you want to proceed at the beginning of the discussion . . . *(Go to 12)*

If you want to proceed to the Z test for:

1. One sample in which the population variance is known . . . *(Go to 12)*
2. One sample in which the population proportion is known . . . *(Go to 55)*
3. Two independent samples for proportions . . . *(Go to 66)*
4. Two independent samples for $\overline{X}$ . . . *(Go to 80)*

---

12 Probably the simplest kind of Z test is the single sample test, which compares a mean ($\overline{X}$) of one sample to the known mean of the population. The example below provides information about a hypothesis test that involves such a test statistic.

A psychologist is interested in checking on the adequacy of the sampling procedures used to obtain a sample of students from the university for one of her experiments. She suspects that students with higher grade point averages (G.P.A.'s) than is the norm for the university may have been selected for sampling in a biased proportion; i.e., they were given greater probability of appearing in the sample. She obtains from the university data that can provide information on the total population (i.e., all students) and finds that the university-wide G.P.A. is 3.7 with a standard deviation of 0.7. The sample size includes 100 students, supposedly randomly selected. The mean G.P.A. of this sample is found to be 3.9. (It is suggested that you copy these statistics on a piece of paper, since they will be referred to frequently.)

You can tell that a Z test, rather than a t test is appropriate because:

A. The population parameter (μ) is known
B. The size of the sample is 100

(A, *go to* 13)
(B, *go to* 14)

---

13 The fact that the population parameter (μ) is known does not necessarily suggest that a Z test would be preferred over the t test. The size of the sample being 100 is the major decision factor in choosing the Z over the t test because the Z test is primarily used with fairly large samples ($N > 30$).

(*Go to* 15)

---

14 A sample size of 100 is the criterion for selecting a Z test over the t test. The t test is used primarily for samples of 30 or less ($N \leqslant 30$).

(*Go to* 15)

---

15 The question to be tested in the case of the psychologist is whether she has reason to suspect a biased sample. The hypothesis, when stated in null form, would be:

A. The sample was selected randomly
B. The university-provided data are correct
C. There were significantly more high-G.P.A. students in the sample than in the university population

(A, *go to* 16)
(B, *go to* 17)
(C, *go to* 18)

---

16 The null hypothesis, H_0, is: The sample was selected randomly; i.e., there is no difference between the sample $\overline{X}$ and the population μ. Your answer is correct. (*Go to* 19)

17 The data supplied by the university about the population must be assumed to be correct. The psychologist has no alternative. The primary interest is with the *sampling procedure*. Therefore the null hypothesis would be that the sample was selected randomly; i.e., there is no difference between the sample $\overline{X}$ and the population μ. The university provided G.P.A. data about the population that will assist in determining whether the sample was representative of the population (unbiased).

(*Go to* 19)

18 Your null hypothesis, H_0, is: There were significantly more high-G.P.A. students in the sample than in the university population. This would be an alternative or research hypothesis because you are postulating a specific direction for the relationship in the sampling procedure. The correct answer is that H_0 is: The sample was selected randomly; i.e., there is no difference between the sample $\overline{X}$ and the population μ. The interest is in rejecting this in favor of the alternative hypothesis.

(*Go to* 19)

19 In testing a statistical hypothesis such as the one regarding the psychologist's sample, before computing the test statistic and making a decision, you would also need to:

A. Determine the credibility of the university's data
B. Obtain a sampling distribution and choose a significance level
C. Choose a significance level

(A, *go to* 20)
(B, *go to* 21)
(C, *go to* 22)

20 Determining the credibility of the university G.P.A. data is not correct, since you have to *assume* that they are correct. You should recall that the steps in doing a hypothesis test are to (1) determine the assumptions (hypotheses, levels of measurement, etc.), (2) obtain the sampling distribution, (3) select a significance level and critical region, (4) compute test statistic, and (5) make a decision.

You have already determined the hypothesis, so the sampling distribution and significance level must also be identified before the test statistic can be computed. (*Go to* 23)

---

21 The significance level and sampling distribution must yet be identified before computing the statistic and making a decision. Your answer is correct.

(*Go to* 23)

---

22 Your answer is partially correct; choosing a significance level is a necessary step. However, the sampling distribution must also be obtained or chosen so that the sample statistic can be properly compared to the population parameter in terms of the probability for events (the statistic) occurring in such a fashion. If you are still hazy on the role of the sampling distribution and significance level, it might be wise for you to review the appropriate parts of chapter 8.

(*Go to* 23)

---

23 Since the central limit theorem indicates that for large sample sizes the sampling distribution of the sample means is normal or approximately normal, we can use the *normal table* for our sampling distribution. (See Appendix A, Table 2.) The normal table will be used frequently from now on. It is important to realize that this table has been computed using probability theory. It represents the probability of all possible outcomes stated in terms of areas under the normal curve.

Choosing a significance level, of course, depends on the risks that you are willing to take of making a Type I or II error. Suppose .05 is chosen ($\alpha = .05$). The only thing left before doing the test is to decide the direction for the test.

For the psychologist's sampling problem, the situation requires a:

A. One-tailed test

B. Two-tailed test

(A, *go to* 24)

(B, *go to* 25)

---

24 The one-tailed test is desired because the data indicated the sample mean to be greater than the population mean, and thus our interest is only in the direction of bias toward a high G.P.A.

(*Go to* 26)

---

25 A one-tailed test is more appropriate because the psychologist suspected over-sampling of students with higher G.P.A.'s. A result less than the population mean of 3.7 would certainly not show sampling bias for higher G.P.A.'s. Interest is in only one direction. Therefore a one-tailed test will be used.

(*Go to* 26)

---

26 Given the .05 significance level and a one-tailed test, the *value* of the critical region must be determined from the normal table. The normal table tells you the area under the normal curve between the mean (which is standardized at zero on the curve) and any ordinate at some distance from the mean, in terms of *standard normal scores*. The standard deviation (or Z values) of the normal curve in the table is set at 1.0.

Thus, if you know the standard deviation for the normal curve, you can use Table 2 to:

A. Find the height or frequency of occurrences relative to the Y axis of the plotted curve
B. Calculate the appropriate significance level for a one-tailed test
C. Determine the area under the curve that corresponds to that value

(A, *go to* 27)
(B, *go to* 28)
(C, *go to* 29)

---

27 The normal table presented in Table 2 will not tell you the frequency of occurrences relative to the Y axis of the normal curve. Since there is no reference to the scale used in the Y axis, the normal table has no way of indicating such information.

The correct answer is that the table will tell you the area under the curve that corresponds to the value of the critical region which you set. Inversely, the table will tell how many standard deviations (standard errors of the statistic) encompass an area when that area is already known.

(*Go to* 30)

---

28 The normal table will not tell you the appropriateness of a level of significance. You have already decided this level and have based your decision on the risks for making Type I and Type II errors.

Most appropriately, your answer should have been that the normal table will tell you the area under the curve that *corresponds* to the value of the critical region that you have already set. Inversely, the table will tell you how many standard deviations or, more appropriately, standard errors of the statistic, encompass an area when you already know that area.

(*Go to* 30)

---

29 The table will tell you the area under the normal curve that corresponds to the value of the critical region. Inversely, the table will indicate how many standard deviations (standard errors of the statistic) encompass an area when that area is already known.

(*Go to* 30)

---

30 Table 2 represents areas under the normal curve and contains three items of information:

1. Column (A) represents the number of normalized (standardized) standard deviations you are away from the mean. This is called a Z score.
2. Column (B) represents the area under the normal curve between the mean and any particular Z score.
3. Column (C) represents the area left in the curve beyond any given Z score. Since 50% of the curve is found on one side of the normal curve, column (B) + column (C) should always equal .5000, or half of the area under the curve.

For example, 1.60 standard deviations from the mean contains .4452 of the curve. A 1.64 standard normal score would contain what proportion of the curve?

A. .4495
B. .0505

(A, *go to* 31)
(B, *go to* 32)

---

31 The answer you chose was .4495 and represents the area between the mean and the Z score of 1.64. The value was found by locating a Z score of 1.64, which corresponds to 1.64 standard errors of the estimate away from the mean. The area beyond the Z of 1.64 represents .0505, or 5.05% of the curve.

(*Go to* 33)

---

32 The area that you chose is .0505, or 5.05% of the curve, and represents the area of the curve beyond the Z score of 1.64. The correct answer is .4495, which represents the area between the mean and the Z score of 1.64. That Z score also represents 1.64 standard errors of the estimate away from the mean.

(*Go to* 33)

---

33 It is crucial for you to realize that the normal curve is a symmetrical distribution. Therefore:

A. The table of areas under the normal curve must be halved to determine the number of standard deviations from the mean

B. The two halves of the curve are mirror images of each other

(A, *go to* 34)
(B, *go to* 37)

---

34 The table of areas under the normal curve must not be halved to determine the standard deviations from the mean. Being symmetrical implies only that the two halves of the curve are mirror images of each other and that what is true of the right side of the curve is also true of the left. Therefore the area for 1.65 standard deviations for the left side or one direction would be the same area for the other direction of deviation from the mean. The area on both sides of the mean, i.e., the area including both directions of a standard deviation of 1.65, would be:

A. .9010

B. .2252

(A, *go to* 35)
(B, *go to* 36)

---

35 The normal table indicates that you will find an area of .4505 for a 1.65 $\sigma_{\bar{X}}$. This must then be doubled to obtain the area for both directions, so your answer of .9010 is correct.

(*Go to* 38)

---

36 You halved the area of .4505 to obtain this answer; it is not correct. To calculate the area for both directions of a $\sigma_{\bar{X}}$ of 1.65 it is necessary to locate the area for one direction in the normal table and then double it (i.e., $.4505 \times 2 = .9010$). You must remember that the normal curve (represented by the normal table) is perfectly symmetrical.

(*Go to* 38)

---

37 The fact that the normal curve is symmetrical implies that the two halves of the curve are mirror images of each other and that what is true of the right-hand side of the curve is also true of the left.

The area on both sides of the mean, i.e., the area including both directions of a standard deviation of 1.65 would be:

A. .9010

B. .2252

(A, *go to* 35)
(B, *go to* 36)

---

38 Let us say that you are interested in finding how many normalized standard deviations you are above the mean ($\mu + \sigma_{\bar{x}}$) when you are concerned with a critical region under the normal curve that is equal to .01 ($\alpha = .01$). In other words, you want to know how many standard deviations you must move above the mean to have .01 of the area under the normal distribution fall above that point on a normal curve. You know that the total area under the curve is equal to unity (area = 1.0000). Therefore the half of the curve containing the critical region is equal to .5000. Since .01 of the curve is in the critical region, it is necessary to subtract that area from half of the normal curve: .5000 − .01 = .4900. That is to say, .4900 of the curve lies in the area between the mean and the point indicated by $\alpha = .01$.

Determine from Table 2 the number of standard deviation units that are represented by an area under the curve beyond the Z score and equal to .01.

When you have done this . . .

(*Go to* 39)

---

39 The number of normalized standard deviation units ($\sigma_{\bar{x}}$ in this problem) that represent .01 of the normal distribution is 2.33. (It is really between 2.32 and 2.33, but extreme accuracy for our purposes is not that necessary; it is appropriate to select 2.33, the nearest value.) In other words, if you were to move 2.33 standard deviation units out from the mean on the normal curve, you would contain .4900 of the curve in that area and have the desired .01 beyond the Z of 2.33. Fig. 10-1 depicts this relationship.

(*Go to* 40)

FIG. 10-1

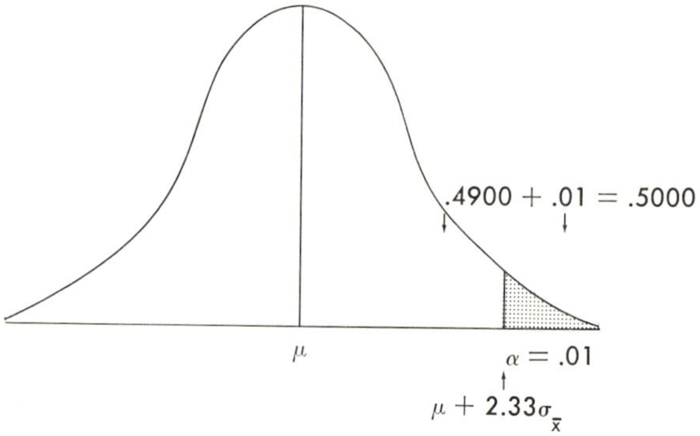

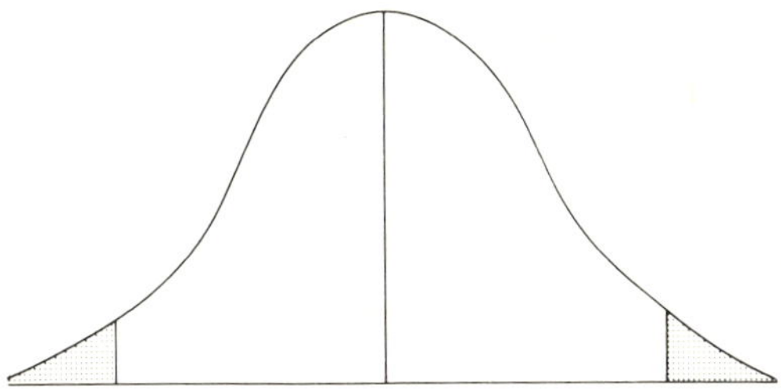

FIG. 10-2

---

40 Not all tests of statistical significance will deal with only one-tailed tests as with the psychologist's research. Therefore it will be necessary to learn how to calculate the number of normalized standard deviation units from the mean for a two-tailed test as well. The process is identical to that for a one-tailed test; the only difference is that you are now dealing with the probability of making a Type I error (α) divided into two equal parts.

Suppose you have $\alpha = .01$ for a two-tailed test. The size of each critical region must be .01/2, or .005. Or, .005 of the normal curve lies beyond the Z score. The corresponding Z is between 2.57 and 2.58 units away from the mean. (Refer to Table 2.) For convenience let us use 2.57 as the value. So for a two-tailed test with $\alpha = .01$, the test statistic that you calculate must have a value of 2.57 or greater if you are to reject the null hypothesis. Fig. 10-2 depicts the normal curve where $\alpha = .01$ for a two-tailed test.

(*Go to* 41)

---

41 Returning now to the psychologist with the sampling problem, the probability of committing a Type I error is set at $\alpha = .05$ for a one-tailed test. (Remember that she was able to predict direction.) How many standard deviation units does this represent on the normal table? Calculate your answer and then . . .

(*Go to* 42)

---

42 Your answer should have been determined by looking in column (C) of Table 2 to find the area beyond Z that corresponds to .05 of the normal curve. The $\sigma_{\bar{X}}$ is somewhere between .0505 and .0495, which corresponds to a Z of 1.64 and 1.65, respectively. For convenience we chose

Continued.

.4505 as the area between μ and the beginning of the critical region, thus arriving at $1.65\sigma_{\bar{X}}$.

In other words, the value of the test statistic must be 1.65 or greater to reject the H_0 with the risk of 5% of the time making a Type I error.

It is now appropriate to compute the Z statistic that will tell the psychologist if the sample mean G.P.A. of 3.9 was within the area of the sampling distribution (normal curve) from μ to $\mu + 1.65\sigma_{\bar{X}}$ or in the critical region, which is beyond $\mu + 1.65\sigma_{\bar{X}}$. The formula for finding the Z statistic when σ (the standard deviation of the population) is known and N is large ($N > 30$) is

$$Z = \frac{\bar{X} - \mu}{\sigma/\sqrt{N}}$$

where

$\bar{X}$ = the mean obtained in the sample
μ = the mean from the population
σ = the standard deviation of the population
N = the size of the sample

The test statistic, Z, represents the difference between the sample mean and the population mean (which is what you are attempting to ascertain) controlled for the standard error of the sampling distribution for the mean.

Another way of saying this is that you are interested in knowing whether the difference between the sample and population characteristics is extreme enough in terms of the shape of the sampling distribution to warrant rejecting a null hypothesis that says they are really the same. The null hypothesis, in this case, would say that there is equality between the

A. Sampling distribution and the sample distribution
B. Standard deviations of the sample and population
C. Means of the sample and population

(A, *go to* 43)
(B, *go to* 44)
(C, *go to* 45)

---

43 There is no logical or useful way of comparing a *sample* distribution, which distributes a *variable* in one sample (e.g., grade-point averages of the 100 students), and a *sampling* distribution, which distributes a *statistic* over many samples (e.g., mean grade-point averages of many samples of students).

(*Go to* 45)

---

44 The null hypothesis will state nothing about comparing the standard deviations between sample and population. The only measures of dispersion available to you are the standard deviation of the population and the standard error of the statistic.

(*Go to* 45)

45 The Z test for the difference between means when the population is known will have a null hypothesis that says $\overline{X} = \mu$, or means of the sample and population are equal.

With your understanding of the null hypothesis, it is now possible to plug the data from the psychologist's problem into the formula, and test your null hypothesis. Recall that the psychologist had a sample of 100 students with a mean G.P.A. of 3.9; the university data indicates that the mean G.P.A. for all students is 3.7 with a standard deviation of 0.7.

$$Z = \frac{\overline{X} - \mu}{\sigma/\sqrt{N}} = \frac{3.9 - 3.7}{0.7/\sqrt{100}} = \frac{0.2}{0.7/10} = \frac{0.2}{0.07} = 2.86$$

Thus the mean G.P.A. from the sample falls in the critical region ($\mu + 1.65\sigma_{\overline{X}}$), since $2.86 > 1.65$. That is to say, we can reject the H_0 of no difference between the sample $\overline{X}$ and the population μ, with the knowledge that 5% of the time ($\alpha = .05$) we will be committing a Type I error. Fig. 10-3 depicts the relationship between the computed Z of 2.86 and the critical region beginning at Z = 1.65.

(*Go to* 46)

FIG. 10-3

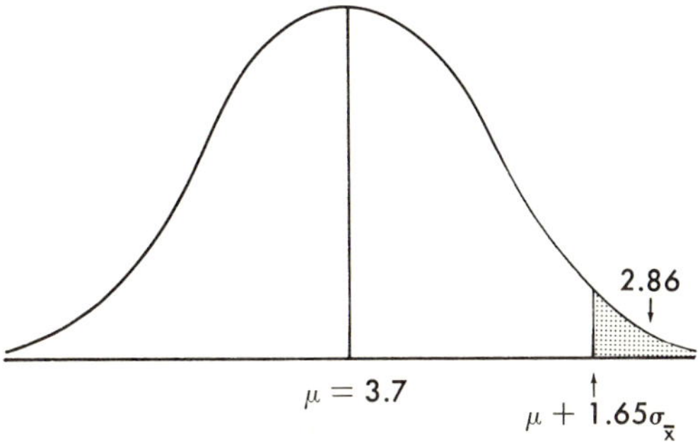

46 Going this far with the test for statistical significance is only half of the task. You know that the difference between the sample and population means are statistically significant, but what does this indicate for the problem at hand and the psychologist?

Continued.

It was hypothesized that students with high G.P.A.'s were oversampled (i.e., biased sampling procedures were used). Based on the fact that the test statistic fell in the critical region, we can conclude that in fact this was *probably* the case. We reject the null hypothesis of no difference between the two means ($\overline{X}$ and μ) and conclude that the alternative hypothesis is correct. We can say that only 5% of the time or less ($\alpha = .05$) would chance variation have produced the observed difference between the sample mean (3.9) and the population or university mean (3.7) if the null hypothesis of no difference were true. Our result is too rare to be chance, and therefore it must be due to the sample being different from the population. We would probably not have a good deal of confidence in the sample data as being representative of:

A. Sample data from another, similar experiment
B. The population characteristics

(A, *go to* 47)
(B, *go to* 48)

---

47 The test statistic, in this case, showed that the sample was not representative of the population characteristics. If G.P.A. were critical to her experiment, the psychologist would probably want to choose another sample. It might be true that the data from a similar experiment would not be the same as that obtained from this sampling (remember that means vary from sample to sample), but this kind of conclusion could not be made from the statistical analysis in this case. Your answer was incorrect.

(*Go to* 49)

---

48 The sample data should not be taken as being representative of the population characteristic. If G.P.A. were critical to her experiment, the psychologist would probably want to choose another sample.

(*Go to* 49)

---

49 Notice this first Z statistic required that:

1. Population mean (μ) and standard deviation (σ) be known
2. Only one sample was used, and it was compared to the population, *not* to another sample
3. $N > 30$

When actually doing research you will need to decide on which statistic is appropriate. The information listed above is necessary for use of this particular Z test.

Another example may be useful before going on. Write down the pertinent information from the example below on a separate sheet of paper.

A physician has obtained a random sample of 125 hypertensive patients' medical charts from the year's population of clinic patients. He knows that the mean blood cholesterol for the clinic population is 250 mg (per 100 ml), with a standard deviation $\sigma = 75$ mg (per 100 ml), whereas the mean blood cholesterol of his sample is 263 mg (per 100 ml). Given the large standard deviation and his unsureness of the direction of the difference between the sample and population means, he wants to test if there is a statistically significant difference between the sample and population mean at the .01 level of significance.

This example differs from the previous one only in that a two-tailed test is used, although it is not necessarily required. Therefore the critical value of Z obtained from the normal table indicating whether the *computed* Z statistic exceeds a certain level of probability must be obtained differently. With a two-tailed test there is a critical region at both ends of the curve. The significance level was set at .01. In this problem .01 must be divided by 2 (i.e., .005), and the Z corresponding to that area must be found. The result, the Z corresponding to the proportion of area under the curve beyond the Z is used as the critical Z value.

What is the Z value for each tail of the distribution?

Make your decision and then . . . (*Go to* 50)

---

50 The value of Z from the normal table would be $\alpha = .01/2 = .005$; .005 of the curve begins at a Z score of 2.57 or 2.58. The calculated Z score must exceed 2.57 to be rejected.

What would be the Z for a two-tailed test when the significance level was .05? Locate the Z in the normal table before going on. (*Go to* 51)

---

51 Alpha = .05/2 = .025; .025 corresponds to a Z = 1.96. Or 1.96 standard errors of the estimate away from the mean represents the beginning of the area under the curve of .025.

Going back to the physician, we can now proceed to compute the Z statistic. The formula, again, is:

$$Z = \frac{\overline{X} - \mu}{\sigma/\sqrt{N}}$$

Calculate the test statistic and decide whether or not you should:

A. Reject the H_0 (A, *go to* 52)

B. Fail to reject H_0 (B, *go to* 53)

---

52 The computation for this test of statistical significance should be as follows:

$$Z = \frac{263 - 250}{75/\sqrt{125}} = \frac{13}{75/11.2} = \frac{13}{6.7} = 1.94$$

The fact that 1.94 does not exceed 2.57 (the necessary Z score for $\alpha = .01$ in a two-tailed test), means that you cannot reject the null hypothesis ($1.94 < 2.57$). Your calculated Z statistic does not fall in the critical region.

(Go to 54)

53 Failing to reject the null hypothesis is the correct answer. The calculation of Z ($\frac{263 - 250}{75/\sqrt{125}} = \frac{13}{75/11.2} = \frac{13}{6.7} = 1.94$) results in a value that is less than the Z value corresponding to $\alpha = .01$ for a two-tailed test of 2.57.

(Go to 54)

54 Based on the result of your test statistic, the physician can be assured that the 125 hypertensive patients in his random sample do not differ significantly from the clinic population in terms of blood cholesterol. In terms of the variable "cholesterol," he does have a random sample of hypertensive patients from the clinic population.

(Go to 55)

55 A Z test can also be used to cover hypothesis tests involving proportions whenever N is fairly large (i.e., $N > 30$). The following example illustrates a problem using proportions.

An investigation is made on an advertising claim that 70% of the school systems in the United States are now using a given learning module for arithmetic in the first grade. After taking a random sample of 50 systems, it is discovered that only 60% are using the module; i.e., a proportion of 0.60. Was the advertising fraudulent? There was a difference of 0.10 between the claim of the advertisers and the evidence of the investigators. Perhaps the difference is due to sampling error. How likely is it that we would obtain a proportion as low as 0.60 if the "true" (population) proportion is 0.70? If it is unlikely that we could get a proportion of 0.60, we have statistical evidence that the advertising claim is fraudulent.

The first step in analyzing this situation would be to determine the *standard error* of the proportion (standard deviation of a sampling distribution made up of proportions). We would then be in a position to make statements of probability about where any sample would fall. The formula for finding σ_{prop} is

$$\sigma_{prop} = \sqrt{\frac{p_u q_u}{N}}$$

where

p_u = the proportion occurring in the population (or universe)
q_u = the proportion not occurring in the population, or $q_u = 1 - p_u$
N = sample size

What would be the standard error of the proportion for the advertising claim made above?

After you have finished your calculation . . . (*Go to* 56)

---

56 The standard error of the proportion would be 0.065. This was found by

$$\sigma_{prop} = \sqrt{\frac{0.70\ (0.30)}{50}} = \sqrt{\frac{0.21}{50}} = \sqrt{0.0042} = 0.065$$

where

0.7 = the advertiser's claim of proportion (hypothetically true for the population)
0.3 = $1 - p_u$
50 = size of the sample made by the investigators

Given the standard error of the proportion (0.065), one could easily locate the sample proportion with relation to the hypothetically true population proportion on a normal curve to determine whether the sampling error of the true proportion is large enough for us to expect the obtained proportion. Fig. 10-4 (p. 196) demonstrates this relationship.

From your knowledge of the normal curve and the concept of Type I (α) errors, the location of the sample proportion, 0.60, is a:

A. Likely proportion
B. Improbable proportion

(A, *go to* 57)
(B, *go to* 58)

---

57 This visual test should suggest that it is feasible to get a sample proportion of 0.60 if the true proportion is 0.70. The 0.60 falls between the first and second standard errors, and not in the extreme tail of the distribution. To be certain, however, it is necessary to apply a test of statistical significance to the problem. (*Go to* 59)

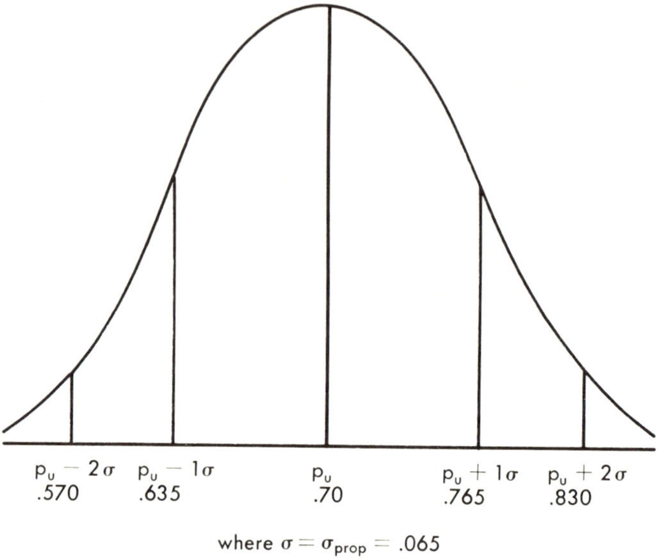

FIG. 10-4

---

58 A proportion of 0.60 has a rather high probability of occurrence when it is between the first and second standard errors. The sample proportion does not fall near the potential critical regions or extreme ends of the curve. Therefore it is feasible to get a sample proportion of 0.60 if the true proportion is 0.70. To be certain, however, it is necessary to apply a test of statistical significance to the problem. (*Go to* 59)

---

59 Note that we have used the normal curve for our sampling distribution. When N is large, proportions, as are means, will be approximately normal in distribution. In terms of actually testing a hypothesis, it is better to figure the amount by which the obtained proportion, or p_s, the proportion in the sample, deviates from the hypothetical population proportion and to make exact statements of probability. The formula for this Z statistic is

$$Z_{prop} = \frac{p_s - p_u}{\sigma_{prop}}$$

where

p_s = proportion of the sample
p_u = true proportion in the population
$\sigma_{prop} = \sqrt{\frac{p_u q_u}{N}}$

Establish a null and an alternative hypothesis for the advertising claim concerning first-grade arithmetic modules and then test the null hypothesis with the formula for Z_{prop}, using $\alpha = .05$. When you have established your H_0 and H_A, calculated the test statistic, and made your decision . . . (*Go to* 60)

---

60 H_0: There is no difference between the population proportion (p_u) and the sample proportion (p_s); or, $p_u = p_s$.

H_A: p_s is less than p_u; or the advertisers overestimated the extent of use (a one-tailed hypothesis).

Z for $\alpha = .05$ from Table 2, in one tail, is found from the area beyond Z of .0500, or $Z_{.05} = 1.65$ units.

$$Z = \frac{p_s - p_u}{\sigma_{prop}} = \frac{0.60 - 0.70}{0.065} = 1.54 \text{ (you can ignore the sign)}$$

Because the calculated Z value (1.54) is less than the Z value from Table 2 (1.65), you have no information or reason to reject the null hypothesis. In other words, there is no reason to believe that the advertisers misrepresented their results, and you can conclude that the difference between the sample proportion (p_s) of 0.60 and the population proportion (p_u) of 0.70 was only a chance occurrence due to sampling and not a "real" phenomenon.

If you examine Table 2, you will see that 1.54 standard error units = .4382, or this value occurs approximately 44% of the time.

If you want to try another example of this type of test . . . (*Go to* 61)
If you think that you adequately understand this test . . . (*Go to* 66)

---

61 You are interested in evaluating the administration of a particular prereading program in a Head Start center. Specifically, you want to determine if the prereading program results for a particular center matched national standards for the program. You have drawn a random sample of 95 cases from the files of the center. The percentage of "successes" is found to be 63% as compared with the standard of 68%. Can you conclude that this center's program was below the national standard?

1. Establish an H_0.
2. Establish an H_A.
3. Determine the critical value of Z (the number of standard error of the proportion units) at $\alpha = .02$ from the appropriate statistical table.
4. Calculate the test statistic.
5. Draw your conclusion.

When you have completed these steps . . . (*Go to* 62)

---

62 Your null and alternative hypotheses should read approximately as follows:

H_0: There is no difference between the national standard and the sample findings, or $p_u = p_s$.

H_A: The sample findings are below those of the national standard, or $p_s < p_u$.

Using $\alpha = .02$ for a one-tailed test results in a normal standard error of the estimate, or a value for the critical region, of 2.05, or $Z_{.02} = 2.05$. The value of the test statistic must be greater than 2.05 to reject H_0 at $\alpha = .02$.

The appropriate test statistic is that of a one-sample proportion Z test when the population proportion is known, or:

$$Z = \frac{p_s - p_u}{\sigma_{prop}}$$

Where:

$$\sigma_{prop} = \sqrt{\frac{p_u q_u}{N}}$$

$$q_u = 1 - p_u$$

Using the data from the Head Start center above results in the following test statistic value:

$$Z = \frac{0.63 - 0.68}{\sqrt{0.68(0.32)/95}} = \frac{0.05}{\sqrt{0.2176/95}} = \frac{0.05}{\sqrt{0.0023}} = \frac{0.05}{0.048} = 1.04$$

Should you:

A. Fail to reject H_0?

B. Reject H_0?

(A, *go to* 63)
(B, *go to* 64)

---

63 Since your test statistic, 1.04, is less than the needed $Z_{.02}$ value of 2.05, you should fail to reject the null hypothesis.

Provide substantive meaning for the problem and then . . .

(*Go to* 65)

---

64 You cannot reject the null hypothesis. Remember that the test statistic must be greater than the Z to reject. Since your test statistic was 1.04, which is less than $Z_{.02}$ of 2.05, you have no basis for rejecting the null hypothesis.

Provide substantive meaning for the problem and then . . .

(*Go to* 65)

65 There is no statistical reason to suspect that because p_s was less than p_u, this difference was "real." It may have been attributed to the sampling variation, and therefore you can conclude that there is no difference between this Head Start center and the national norm for a successful prereading program. That is, the percentage "successful" in this center (63%) does not differ significantly from the national norm of 68%.

(*Go to* 66)

---

66 There is yet a second kind of Z statistic for proportions, and it is used in cases involving two independent samples.

Suppose a comparison of learning patterns is made between college students in the humanities (sample 1) and students whose education was oriented more toward basic science and research (sample 2). Let us assume that the researcher suspects that when given an option, basic science students will be more inclined to select "active" rather than "passive" listening and reading forms of study and learning activities. Among a random sample of 150 basic science students, it is found that 57% list "active" forms of learning activities as their choice. Of the humanities students (also selected randomly), 46% of the 120 students list "active" activities as their preferred learning experiences. Is there a significant difference at the .05 level?

A comparison must be made of two independent random samples with respect to the proportion of "active" learning choices. What we are interested in here is whether the two types of educational orientation (humanities and basic science) have the same or different population proportions regarding "active" learning.

What will be the H_0 and H_A? Write them down and then . . .

(*Go to* 67)

---

67 H_0: There is no difference between the proportion of humanities students (p_{u_1}) and the proportion of science students (p_{u_2}) with regard to their selection of "active" learning experiences; or $p_{u_1} = p_{u_2}$

H_A: The basic science students prefer "active" learning experiences more often than do the humanities students; or $p_{u_1} < p_{u_2}$

You will notice that the subscripts u_1 and u_2 are used in the null and alternative hypotheses to represent the two types of educational orientation. We use subscripts representing the population (or universe) rather than the sample in the null and alternative hypotheses because, as you will recall from the discussion of the null hypothesis, what we are really interested in when dealing with inferential statistics are the values for

Continued.

the population and not the sample per se. However, the only data that we have are for the samples, and therefore in the actual calculation of the test statistics we use the sample statistics to represent the population parameters.

The formula for finding the Z statistic when comparing two independent samples with proportions is:

$$Z = \frac{p_{s_1} - p_{s_2}}{\hat{\sigma}_{p_{s_1} - p_{s_2}}}$$

The denominator, of course, is a new notation and is called the standard error for a difference of proportions. An estimate of variance ($\hat{\sigma}$) is made because there is no known or hypothetically true value of σ as in the previous cases. Furthermore, rather than estimate the variance in each sample with different N's, we pool the values and use a single estimated proportion ($\hat{p}_u$). The computation of this variance is made by using the formula:

$$\hat{\sigma}_{p_{s_1} - p_{s_2}} = \sqrt{\hat{p}_u \hat{q}_u} \sqrt{\frac{N_1 + N_2}{N_1 N_2}}$$

Now the question is, how does one find the pooled estimate ($\hat{p}_u$) for the population? The Z statistic cannot be calculated until the estimated standard error for the difference of proportions, $\hat{\sigma}_{p_{s_1} - p_{s_2}}$, is determined, and it, in turn, cannot be determined until the pooled estimate for the population proportion ($\hat{p}_u$) is obtained. Therefore one last formula must be introduced, the one for finding $\hat{p}_u$:

$$\hat{p}_u = \frac{N_1 p_{s_1} + N_2 p_{s_2}}{N_1 + N_2}$$

Then $\hat{q}_u$ will be determined by subtracting $\hat{p}_u$ from 1 (i.e., $\hat{q}_u = 1 - \hat{p}_u$). The proportions occurring in the sample are p_{s_1} and p_{s_2}.

Calculate the pooled estimate for the population ($\hat{p}_u$) for the humanities and basic science problem, then . . .

(Go to 68)

---

68 Your calculations should be similar to those that follow:

$$\hat{p}_u = \frac{N_1 p_{s_1} + N_2 p_{s_2}}{N_1 + N_2} = \frac{120(0.46) + 150(0.57)}{120 + 150} = \frac{55.2 + 85.5}{270} = \frac{140.7}{270} = 0.521$$

It is now possible to determine the standard error of the difference of proportions using the pooled estimate that you just calculated. Using the formula for $\hat{\sigma}_{p_{s_1} - p_{s_2}}$ given above, determine what the standard error of the difference of proportions would be for the problem. When you are finished . . .

(Go to 69)

---

69 If $\hat{p}_u = 0.521$, then:

$$\hat{q}_u = 1 - \hat{p}_u = 1.000 - 0.521 = 0.479$$

Then substituting in the formula:

$$\hat{\sigma}_{p_{s_1} - p_{s_2}} = \sqrt{\hat{p}_u\hat{q}_u}\sqrt{\frac{N_1 + N_2}{N_1N_2}} = \sqrt{0.521(0.479)}\sqrt{\frac{120 + 150}{120(150)}}$$

$$= \sqrt{0.2496}\sqrt{\frac{270}{18{,}000}} = \sqrt{0.2496}\sqrt{0.015} = \sqrt{0.0037} = 0.061$$

Now $\hat{\sigma}_{p_{s_1} - p_{s_2}}$ is known (0.061) and this can be substituted into the Z formula:

$$Z = \frac{p_{s_1} - p_{s_2}}{\hat{\sigma}_{p_{s_1} - p_{s_2}}}$$

Calculate the Z statistic and then . . .

(*Go to* 70)

70 $Z = \dfrac{p_{s_1} - p_{s_2}}{\hat{\sigma}_{p_{s_1} - p_{s_2}}} = \dfrac{0.46 - 0.57}{0.061} = \dfrac{-0.11}{0.061} = -1.80$

If $\alpha = .05$ for a one-tailed test, what does $Z = -1.80$ indicate about the problem?

A. It can be said that there is no difference between the two proportions.

B. The null hypothesis of no difference can be rejected; i.e., there is a difference between the two proportions.

(A, *go to* 71)
(B, *go to* 72)

71 Of the area under the curve, .05 is beyond the Z value of 1.65. Therefore the null hypothesis should be rejected, and you should conclude that there is a difference between the two proportions. Remember that we are only dealing with one tail here, since the hypothesis included a direction of difference. Therefore the Z value of -1.80 would fall in the critical region, and the H_0 would be rejected.

There seems to be a difference between the preferred learning patterns of the two types of students. Students with basic science backgrounds prefer "active" learning more frequently than do those students with humanities backgrounds.

(*Go to* 73)

72 An $\alpha = .05$ for a one-tailed test corresponds to a critical value for Z of 1.65. Therefore the computed Z of −1.80 would fall in the critical region, and the H_0 could, as you correctly concluded, be rejected. It is possible that there is a difference between the preferred learning patterns of the two types of students. Students with basic science backgrounds prefer "active" learning more frequently than do those students with humanities backgrounds. (*Go to* 73)

73 It is found that in a random sample (sample 1) of 200 children from randomly sampled towns with fluoride-treated water supplies, 12% have had two or more cavities during the past year. In a separate random sample (sample 2) of 500 children from randomly selected towns with untreated water supplies, 20% have had two or more cavities in the past year. The difference is 8 percentage points in favor of the children in towns with fluoridated water supplies. Would a difference this great, and in this direction, occur more than 5 times in 100 if fluoridation really made no difference?

Calculate the appropriate test statistic to answer the problem. Be sure to set up a null and an alternative hypothesis.

For the answer to the problem . . .

(*Go to* 74)

For the step-by-step explanation of how to solve the problem . . .

(*Go to* 75)

74 The correct Z is −2.50. If you made an error or wish to go through the calculation sequence . . .

(*Go to* 75)

What decision would you make on the basis of a .05 significance level and a one-tailed test?

A. Fail to reject the null hypothesis.
B. Reject the null hypothesis.

(A, *go to* 76)
(B, *go to* 77)

75 There are four separate operations that you must perform to calculate the Z statistic:

$$1.\quad Z = \frac{p_{s_1} - p_{s_2}}{\hat{\sigma}_{p_{s_1} - p_{s_2}}}$$

$$2.\quad \hat{\sigma}_{p_{s_1} - p_{s_2}} = \sqrt{\hat{p}_u \hat{q}_u}\sqrt{\frac{N_1 + N_2}{N_1 N_2}}$$

3. $\hat{p}_u = \dfrac{N_1 p_{s_1} + N_2 p_{s_2}}{N_1 + N_2}$

4. $\hat{q}_u = 1 - \hat{p}_u$

First, the pooled estimate of the proportion must be calculated:

$$\hat{p}_u = \frac{N_1 p_{s_1} + N_2 p_{s_2}}{N_1 + N_2}$$

Substituting in the formula you get:

$$\hat{p}_u = \frac{200 \times 0.12 + 500 \times 0.20}{200 + 500} = 0.177$$

Then:

$$\hat{q}_u = 1 - \hat{p}_u = 1 - 0.177 = 0.823$$

Next, the standard error of the difference of proportions using the pooled estimate is found with the formula:

$$\hat{\sigma}_{p_{s_1} - p_{s_2}} = \sqrt{\hat{p}_u \hat{q}_u} \sqrt{\frac{N_1 + N_2}{N_1 N_2}} = \sqrt{0.177 \times 0.823} \sqrt{\frac{200 + 500}{200 \times 500}}$$

$$= \sqrt{0.1457} \sqrt{\frac{700}{100{,}000}} = \sqrt{0.1457}\ \sqrt{0.007}$$

$$= \sqrt{0.001} = 0.032$$

The standard error of the difference of proportion is 0.032. The last calculation is for the Z formula:

$$Z = \frac{p_{s_1} - p_{s_2}}{\hat{\sigma}_{p_{s_1} - p_{s_2}}} = \frac{0.12 - 0.20}{0.032} = -2.50$$

What decision would you make about the null hypothesis if the significance level were .05 and if a one-tailed test were used?

A. Fail to reject the null hypothesis.
B. Reject the null hypothesis.

(A, *go to* 76)
(B, *go to* 77)

---

76 The Z score for the area beyond Z, to the left of the curve and corresponding to .05, is −1.65. Since the calculated Z = −2.50, which is greater than −1.65, the null hypothesis can be rejected. In other words, it is fairly safe (95% of the time) to conclude from the data that the addition of fluoride to the water has a statistically significant effect on reducing dental caries.

(*Go to* 78)

---

77 Since Z = −2.50, which is greater than −1.65 corresponding to the area of the curve beyond $Z_{.05}$, the null hypothesis can be rejected. In other words, it is fairly safe (95% of the time) to conclude that the addition of fluoride to the water has a statistically significant effect on reducing dental caries.

(*Go to* 78)

78 Another interpretation can be made of the calculated Z statistic or of any test of statistical significance. Some statisticians in substantive disciplines such as sociology or psychology would argue that it is really not necessary to establish, before the statistical procedure, the exact level of probability of committing a Type I (α) error that one is willing to accept. By not establishing the probability of making a Type I (α) error, it is possible to state the exact probability.

For the fluoride example a Z value of −2.50 was obtained. If you look at Table 2, you can determine that this corresponds to an area beyond Z of .0062. This represents a small probability of making an error.

Had you obtained a Z = 1.22, what would be the probability of this occurring by chance? Make your decision and then . . .

(*Go to* 79)

79 Z = 1.22 corresponds to an area of .1112 beyond Z. Since $p < .12$, there is a fairly strong possibility that this finding might have happened by chance. In fact, in the long run, 12% of the outcomes in a given direction would have probably happened by chance. This does not lead to a great deal of confidence in one's results.

Note above that we have selected $p < .12$ to report for an exact probability level rather than $p = .1112$. We do this because it is traditionally acceptable to make the statement that the probability of such a finding occurring by chance is somewhat less than (p<) a given number. Therefore always choose some number slightly larger than the actual table number to report your findings. Incidentally, you might have noted your findings in the following manner: $.1111 < p < .1113$.

(*Go to* 80)

80 Frequently we are interested in testing whether there is a statistically significant difference between two groups that are independently sampled. In this situation, the researcher has known sample means from each of these groups. The following problem is an example of this situation.

The mean number of trials required by a sample of 619 white subjects to learn a list of ten nonsense symbols is 6.74 with a standard deviation of 2.56. The mean number of trials for a group of 130 black subjects is 7.86 with a standard deviation of 2.50. Does it take black subjects, on the average, longer than white subjects to learn nonsense symbols, or is the difference merely due to chance variations inherent in sampling?

Given your knowledge of hypothesis formulation, establish a null and alternative hypothesis for the nonsense symbol problem above. When you are done . . .

(*Go to* 81)

---

81 H_0: There is no difference between blacks and whites in terms of number of trials required to learn nonsense symbols ($\mu_1 = \mu_2$)

H_A: There is a difference in the number of trials required between the two groups ($\mu_1 \neq \mu_2$)

Note that no direction for the alternative hypothesis was established because there is essentially no theory that would allow you to predict the direction in which these results should fall. In this situation, you should be conservative in your selection of the probability of an α error, and therefore a two-tailed test must be selected. If $\alpha = .05$, then $.05/2 = .025$ of the area beyond Z. The Z score needed to achieve significance will be 1.96.

The Z test formula for this situation will be

$$Z = \frac{\overline{X}_1 - \overline{X}_2}{\sqrt{\frac{s_1^2}{N_1} + \frac{s_2^2}{N_2}}}$$

where

s_1^2 = variance (standard deviation squared) of the first sample (whites)
s_2^2 = variance of the second sample (blacks)
$\overline{X}_1$ and $\overline{X}_2$ = means of the two samples
N_1 and N_2 = size of each sample

Using the data supplied in the problem, calculate the test statistic. When you have finished . . .

(*Go to* 82)

---

82 Your substitution and calculation should be as follows:

$$Z = \frac{\overline{X}_1 - \overline{X}_2}{\sqrt{\frac{s_1^2}{N_1} + \frac{s_2^2}{N_2}}} = \frac{6.74 - 7.86}{\sqrt{\frac{(2.56)^2}{619} + \frac{(2.50)^2}{130}}} = \frac{-1.12}{\sqrt{\frac{6.55}{619} + \frac{6.25}{130}}}$$

$$= \frac{-1.12}{\sqrt{0.011 + 0.048}} = \frac{-1.12}{\sqrt{0.059}} = \frac{-1.12}{0.243} = -4.61$$

Reach a conclusion and then . . . (*Go to* 83)

---

83 The result, −4.61, is greater than −1.96 ($Z_{.05}$ for a two-tailed test), and therefore you would reject the null hypothesis and conclude that given a 5% chance of making an error, you are reasonably sure that the population of black subjects is slower than the population of white subjects at learning nonsense symbols. (*Go to* 84)

---

84 The t test is a test that enables a decision about statistical data resulting from a sample with a small N (i.e., $N \leq 30$). Essentially, the t test differs from the Z test by using a more conservative sampling distribution to compensate for the smaller sample size. In all other regards, it is virtually identical to the Z statistic.

The following example poses a problem similar to that using a Z test for a single sample where the population mean is known. However, because N is less than 30, the t test is more appropriate.

A group of 17 persons who are participants in a therapy group were selected for an experiment in alpha wave production on the basis of an expressed "interest" in the phenomenon. They were subsequently given a test measuring their knowledge of the brain's ability to produce alpha waves. The sample test results indicated an $\overline{X} = 84$ and an $s = 16$. The standard level of knowledge for the population of similar persons is a mean of 78. Is it reasonable to assume that the sample was representative of a population in which $\mu = 78$?

Essentially, the test will determine whether:

A. The mean of one sample equals the mean of another sample

B. The mean of one population is equal to the mean of another population

C. The mean of a population is accurately reflected by the mean of a sample

(A, *go to* 85)
(B, *go to* 86)
(C, *go to* 87)

---

85 There is only one sample in the problem, and it has an $\overline{X}$ of 84. The mean of 78 signifies the mean of the population. Therefore we are not concerned with comparing the sample means. The question posed in the problem was whether the sample mean of 84 (which was a sample of only 17 students from that population) deviated significantly so as not to be representative or reflect a population mean of 78.

Dispersion of the population in this case is:

A. Known

B. Unknown

(A, *go to* 88)
(B, *go to* 89)

86 There is only one population mentioned—that of persons taking the standard examination. The other group was a sample drawn from that same population. More correctly, the question posed in the problem was whether the sample mean (of only 17 subjects) deviated significantly so as not to be representative or reflect a population mean of 78.

Dispersion of the population in this case is:

A. Known

B. Unknown

(A, *go to* 88)
(B, *go to* 89)

87 The question posed in the problem is whether the sample mean deviated significantly so as not to be representative or reflect a population mean of 78. Your answer is correct.

Dispersion of the population in this case is:

A. Known

B. Unknown

(A, *go to* 88)
(B, *go to* 89)

88 The deviation which is known is the sample deviation. Population variability is missing. Your answer was not correct. The t test also allows you to estimate the population standard deviation from the sample standard deviation.

Now establish a null and alternative hypothesis (two tails) for the alpha wave experiment. When you have finished . . .

(*Go to* 90)

89 Population variability is missing; only the sample standard deviation is known. Your answer is correct. The t test also allows you to estimate the population standard deviation from the sample standard deviation.

Now establish a null and alternative hypothesis (two tails) for the alpha wave experiment. When you have finished . . .

(*Go to* 90)

90 Your null and alternative hypothesis should be approximately as follows:

H_0: $\overline{X} = \mu$; or there is no difference between the mean of the sample and the population mean on the knowledge test

H_A: $\overline{X} \neq \mu$; or there is a difference between the mean of the sample and the population mean on the test

Because of the theoretical logic of the central limit theorem, one is able to use the following formula to compute a measure of relationship between a sample mean and a population mean

$$t = \frac{\overline{X} - \mu}{s/\sqrt{N - 1}}$$

where

$\overline{X}$ = mean of sample
μ = mean of population
s = standard deviation of sample
N − 1 = the degrees of freedom, where N is the size of the sample

Note that the formula for t is essentially the same as that for Z:

$$\frac{\text{Obtained statistic} - \text{Expected statistic}}{\text{Standard error of statistic}}$$

In the case of the t test, we must estimate the standard error from the data in the sample.

What is the value of t for the alpha wave experiment? When you have finished . . .

(*Go to* 91)

91 Your calculation should be as follows:

$$t = \frac{\overline{X} - \mu}{s/\sqrt{N - 1}} = \frac{84 - 78}{16/\sqrt{17 - 1}} = \frac{6}{16/\sqrt{16}} = \frac{6}{16/4} = \frac{6}{4} = 1.50$$

As with the Z distribution, the t value is relevant only insofar as it is compared to a probability of occurrence, i.e., a sampling distribution. The question asked is what is the probability of occurrence of the difference between observed and expected values, expressed in standard error units, given an approximately normal distribution? The distribution of t is approximately normal, but it deviates farther from normal as N gets smaller. The distribution of t gets flatter as N gets smaller, and thus the standard deviation gets larger. Consequently, a critical value of t will be equal to or larger than a critical value for Z.

Critical values of t can be found in Appendix A, Table 3. The t table is essentially the same as the Z table except that it contains a column (the first one) called "df." Df stands for *degrees of freedom*, a concept that will be left unexplained until you deal with chi-square. However, to determine the t value at a given significance level, df is determined by use of the following formula:

$$df = N - 1$$

For the alpha wave experiment:

$$df = N - 1 = 17 - 1 = 16$$

For this problem, if $\alpha = .05$ for a two-tailed test, the appropriate table value of the t test is found by looking down the column under "level of significance for 2-tailed test," .05. The correct table value for df = 16 is t = 2.120.

From the table it can be seen that the value necessary for rejecting H_0 is:

A. Less than that obtained from the t test for the alpha wave experiment
B. Greater than that obtained from the t test for the alpha wave experiment

(A, *go to* 92)
(B, *go to* 93)

---

92 If you located the .05 significance level column for a two-tailed test and went down the column until you reached the row with a df of 16, the value of t should be 2.120. This is a value greater than that arrived at in the calculation. Thus the mean of the sample does not fall in the area of those outcomes of improbable occurrence for a μ of 78, and therefore:

A. There *probably is a difference* between the sample $\bar{X}$ and the population μ
B. There is *probably no difference* between the sample $\bar{X}$ and the population μ

(A, *go to* 94)
(B, *go to* 95)

---

93 By locating the value for t at the .05 significance level for a two-tailed test and a df of 16, you should find a t of 2.120, which is greater than the calculated t (1.50). Thus the mean of the sample does not fall in the area of those outcomes of improbable occurrence for a μ of 78, and therefore:

A. There *probably is a difference* between the sample $\overline{X}$ and the population μ

B. There is *probably no difference* between the sample $\overline{X}$ and the population μ

(A, *go to* 94)
(B, *go to* 95)

---

94 The calculated t was less than 2.120 and therefore does not fall within the critical region (i.e., the calculated t would have to be 2.120 or greater). You have no basis for rejecting your null hypothesis. Thus it can be said that the mean of the sample probably does reflect the population mean. The therapy group participants are not different from the standard group in their knowledge of the brain's ability to produce alpha waves.

(*Go to* 96)

---

95 Because the calculated t did not equal or exceed 2.120, the mean of the sample was probably representative of the mean of the population. You have no basis for rejecting your null hypothesis. Thus it can be said that the mean of the sample probably does reflect the population mean. The therapy group participants are not different from the standard group in their knowledge of the brain's ability to produce alpha waves.

(*Go to* 96)

---

96 The t test can also be used to compare two independent samples where at least one sample has a small N. The following problem illustrates a situation of this kind. Suppose we test five rats for speed in running an already-learned maze and reward each with a sunflower seed after each successful trial. Suppose further that another group of five is given cerebral stimulation through an implanted electrode after each successful trial. The question is which kind of reward produces the most rapid performance. Reported below are the mean number of seconds taken by each group to complete the maze run as well as their standard deviations.

Sunflower seed	*Cerebral stimulation*
$N_1 = 5$	$N_2 = 5$
$\overline{X}_1 = 19.8$ seconds	$\overline{X}_2 = 15.7$ seconds
$s_1 = 2.8$	$s_2 = 2.4$

Although both N's are equal, this need not necessarily be the case. The df must also be found, and, again, it is dependent on the number of cases in each sample, minus one. However, the formula must be applied to both samples and then added. The formula would be:

$$df = (N_1 - 1) + (N_2 - 1)$$

What would be the degrees of freedom for this problem? When you have calculated the answer . . .

(*Go to* 97)

---

97 $df = (N_1 - 1) + (N_2 - 1) = (5 - 1) + (5 - 1) = (4) + (4) = 8$

What would be the null and alternative hypothesis for the maze-running experiment? Once you have established these . . .

(*Go to* 98)

---

98 H_0: There is no difference in running speed between the sunflower-seed–rewarded rats and the cerebrally stimulated rats, or:

$$(\mu_1 - \mu_2) - (\overline{X}_1 - \overline{X}_2) = 0$$

H_A: There is a difference between the two types of reward in terms of running time in the already-learned maze, or:

$$(\mu_1 - \mu_2) - (\overline{X}_1 - \overline{X}_2) \neq 0$$

The formula for the t test for two independent samples is

$$t = \frac{\overline{X}_1 - \overline{X}_2}{\sqrt{\frac{N_1 s_1^2 + N_2 s_2^2}{N_1 + N_2 - 2}} \sqrt{\frac{N_1 + N_2}{N_1 N_2}}}$$

where

N_1, N_2 = sample size
s_1, s_2 = standard deviation
$\overline{X}_1, \overline{X}_2$ = means

Calculate the t for the problem and then . . .

(*Go to* 99)

---

99 $t = \dfrac{19.8 - 15.7}{\sqrt{\dfrac{5(2.8)^2 + 5(2.4)^2}{5 + 5 - 2}}\sqrt{\dfrac{5 + 5}{5 \times 5}}} = \dfrac{4.1}{\sqrt{\dfrac{5(7.84) + 5(5.76)}{10 - 2}}\sqrt{\dfrac{10}{25}}}$

$$= \frac{4.1}{\sqrt{\dfrac{39.2 + 28.8}{8}}\sqrt{0.4}} = \frac{4.1}{\sqrt{\dfrac{68}{8}}\sqrt{0.4}} = \frac{4.1}{\sqrt{8.5}\sqrt{0.4}} = \frac{4.1}{\sqrt{3.4}} = \frac{4.1}{1.84} = 2.23$$

Interpret your findings and then . . .

(*Go to* 100)

---

100 Using a two-tailed test, since no theory or prior knowledge allows us the less conservative one-tailed test, a $t = 2.23$ with $df = 8$ would have happened by chance less than 10% of the time. That is, with a $df = 8$, a table t value of 1.860 is less than the calculated t of 2.23 and corresponds to a two-tailed test at $\alpha = .10$. On the other hand, the table t of 2.306 corresponding to $\alpha = .05$ for two tails is larger than our calculated t value of 2.23.

Therefore we might reject the H_0 and realize that 10% of the time ($.05 < p < .10$) we are going to be wrong if we imply that there is any difference between the two reward systems. You would probably be safer with $p < .10$ in not rejecting the null hypothesis and concluding that there is no difference between the speed at which a rat runs a maze and the type of reward used.

You should realize that there are other forms of Z and t tests. These generally apply to situations in which the two samples are correlated. Here you have been dealing with two independent samples; a correlated sample is one in which there is a test-retest situation such as found in a before-after experiment. Most standard statistics textbooks cover these situations. Furthermore, you need to have additional knowledge of statistics, such as the correlation coefficient found later in this book, to handle these situations. It is appropriate here to indicate that the generic form of the difference of means and proportions tests has been presented.

Thus far the discussion has dealt only with the Z and t tests. Next we will consider confidence intervals, chi-square, and some guidelines for making decisions about which test to use.

(*Go to* 101)

---

101 Often when we are estimating a population value or parameter based on data from a sample, we simply report our best estimate (e.g., $\bar{X}$ or s). But we recognize that our estimate may be in error; it is based on a sample. So it is common practice to report an interval within which we are reasonably sure the true population value lies. As an illustration, the obstetrician may predict to his patient that delivery will occur about March 10, plus or minus a couple of weeks. He is indicating an interval, derived from his past experience, within which he is reasonably confident of being correct. In general, however, he cannot say exactly how sure he is of his estimate.

The larger our interval the more confident we are that the population parameter will be in the interval. The obstetrician is confident of an interval "March 10 plus or minus 30 days" and not at all confident about "March 10 plus or minus 2 days." The *confidence interval*, based on inferential statistics, allows us to evaluate the exact probability of error.

The confidence intervals discussed here can be obtained because we assume that the sampling distribution of our statistic of interest is approximately normal. The confidence interval depends on the statistic and its standard error. What is a standard error?

A. The probability of a statistic being correct in estimating the population parameter
B. The standard deviation of the sampling distribution of the statistic

(A, *go to* 102)
(B, *go to* 103)

---

102 The standard error of a statistic is the standard deviation of the sampling distribution for the statistic. If from a population you draw 100 samples of the same size (N) and compute a statistic from each sample (e.g., $\overline{X}$), then the distribution of the 100 statistics is called the sampling distribution. In the case of the distribution of sample means ($\overline{X}$'s), the sampling distribution will be normal if the size of each sample is sufficiently large. The standard deviation of the sampling distribution for means for samples of size N is the *standard error of the mean.* It is an *indicator* of the error in our estimates of the true population mean (μ) due to the variations among estimates inherent in sampling.

(*Go to* 103)

---

103 The standard error is the standard deviation of the sampling distribution. Since we are dealing with normal distributions, you should recall from the discussion of the normal curve that about two thirds of our estimates (68.26%) will lie within ± 1 standard deviation (standard error) of our central estimate, and ± 1.96 standard deviations (standard errors) will encompass 95% of our cases. (Refer to Table 2 of the Appendix for this value.) In the normal curve, .0250 of the area lies beyond a $Z = 1.96$. Thus when considering both tails of the sampling distribution, 5% of the area is in the tails, and $Z = 1.96$ will define the limits of 95% of the sample means.

If we knew μ and $\sigma_{\overline{X}}$, we would not need a confidence interval. Generally, however, all we have are $\overline{X}$ and s from one sample. Our *procedure* enables us to use $\overline{X}$ as an estimate of μ and be confident we can establish

Continued.

an interval that will include μ. If we bracket our statistic (here $\overline{X}$, but it could be any statistic) with ± 1.96 standard deviations of the statistic (what we call the standard error), we will encompass our population value 95 times out of 100. Note that the population parameter μ is fixed or "true." It is our sample estimates, $\overline{X}$, that vary, and 95% of the time our *procedure* will produce an interval that includes μ.

From the preceding discussion it can be seen that the confidence interval can be generalized as:

$$\text{Obtained statistic} \pm Z_\alpha(\text{Standard error of obtained statistic})$$

Z_α comes from the normal curve tables (or t tables for small samples) and represents the number of standard error units that we must include to cover the proportion of cases indicated by the given alpha level. When $\alpha = .05$ (our error probability), our confidence level is said to be $1 - .05 = .95$, or 95%. For most cases we can use $Z_{.05} = 1.96$.

Recall that the standard error of the mean is estimated by $\frac{s}{\sqrt{N}}$.

Which of the following is the confidence interval for the mean?

A. $\frac{s}{\sqrt{N}} \pm 1.96\overline{X}$

B. $\overline{X} \pm 1.96\frac{s}{\sqrt{N}}$

C. $\overline{X} \pm 1.96s$

(A, *go to* 104)
(B, *go to* 105)
(C, *go to* 106)

---

104 The model for the confidence interval is:

$$\text{Obtained statistic} \pm Z_\alpha(\text{Standard error of obtained statistic})$$

In a confidence interval for the *mean*, the mean is said to be the obtained statistic that we are interested in. Z_α for $\alpha = .05$ is indeed 1.96.

(*Go to* 105)

---

105 Using the model for the confidence interval

$$\text{Obtained statistic} \pm Z_\alpha(\text{Standard error of obtained statistic})$$

will result in a confidence interval of $\overline{X} \pm 1.96\frac{s}{\sqrt{N}}$.

Williams et al. (1973) collected information on voluntary associations for 380 respondents. They report a mean participation rate of 0.897 with a standard deviation of 1.504. Compute the 95% confidence interval for the mean number of voluntary associations. When you are done . . .

(*Go to* 107)

---

106 For the mean, the estimated standard error is $\frac{s}{\sqrt{N}}$ (the standard deviation of the variable X over the square root of the sample size).

(*Go to* 105)

107 Using the formula and substituting the number regarding voluntary association results in the following:

$$\bar{X} \pm 1.96 \frac{s}{\sqrt{N}} = 0.897 \pm 1.96 \frac{1.504}{\sqrt{380}} =$$

$$0.897 \pm 1.96 \frac{1.504}{19.5} = 0.897 \pm 1.96(0.077) =$$

$$0.897 \pm 0.151 \text{ or } 0.746 \text{ to } 1.048$$

The interpretation of these numbers is that "the 95% confidence interval for the mean of this sample runs from 0.746 to 1.048 voluntary participations." Note that you should *not* say that the probability is .95 that μ (the population mean) lies within the interval. The distinction between these two statements is subtle because we are conceptually or substantively concerned with our sample; statistical inference is concerned with our *procedure*. It is the procedure in which we have confidence. We can say that "95 times out of 100 the confidence interval procedure will produce an interval that will contain the population value."

In summary, remember that to compute the confidence interval for a statistic you must (1) establish the confidence level you wish, (2) determine the appropriate Z_α or t_α (always the equivalent of two tails) and (3) establish the appropriate standard error term. Various standard error terms are given in the discussion on Z and t tests.

(*Go to* 108)

108 Sethi et al. (1973) conducted a random sample survey of Lucknow, India in an attempt to determine the effects of industrialization on the mental health of rural and urban populations. Table 10-1 is abstracted from some of their data. Examine the table and then determine which test would be used if the H_0 was that *all* cells in the table are equal?

A. Z test for proportions
B. t test
C. Neither

(A, *go to* 109)
(B, *go to* 110)
(C, *go to* 111)

TABLE 10-1. Psychiatric diagnoses in Lucknow, India, by place of residence*

Diagnosis	*Place of residence* Rural	Urban	Total
Psychoneurosis (excluding depression)	(a) 14	(b) 112	126
Depression	(c) 4	(d) 36	40
Schizophrenia	(e) 3	(f) 15	18
Mental retardation	(g) 68	(h) 109	177
Total	89	272	361

*Modified from Sethi, B. B., Nathawat, S. S., and Gupta, S. C. 1973. Depression in India. J. Soc. Psych. **91:**3-13 (Oct.).

109 A Z test for proportions cannot be used because the null hypothesis for the Indian data involves the simultaneous comparison of more than two categories of the variables. Although it is possible to reduce each frequency to a proportion, how would you simultaneously compare the proportion of rural psychoneurosis to urban psychoneurosis to rural depression to . . . etc.?

(*Go to* 111)

110 A t test will not work in this situation because you must simultaneously compare several categories of each variable.

(*Go to* 111)

111 Neither test will suffice because the null hypothesis involves a simultaneous comparison of more than two *categories* (four types of diagnosis and two places of residence). The Z test and t test are limited to comparison of two samples or two categories of a variable. Thus another test statistic must be used. This last test, chi-square, will enable the researcher to interrelate *nominal scales* with any number of categories or samples to determine statistical significance. Chi-square also enables a test to be made for differences among three or more samples.

Essentially, chi-square enables an evaluation of whether frequencies that have been *obtained* (obtained frequencies) in a research situation differ significantly from those which would be *expected* (expected frequencies) under certain theoretical assumptions. To illustrate the calculation of chi-square, refer to the *contingency table* (Table 10-2). The subject of the research from which the data were derived is the same as in the previous example. This time there are:

A. Four values (categories) of one variable
B. Two dichotomized variables

(A, *go to* 112)
(B, *go to* 113)

TABLE 10-2. Psychiatric diagnoses in Lucknow, India, by place of residence*

	Place of residence		
Diagnosis	*Rural*	*Urban*	*Total*
Psychoneurosis, depression, or schizophrenia	(a) 21	(b) 163	184
Mental retardation	(c) 68	(d) 109	177
Total	89	272	361

*Modified from Sethi, B. B., Nathawat, S. S., and Gupta, S. C. 1973. Depression in India. J. Soc. Psych. **91**:3-13 (Oct.).

---

112 Your selection of four values of one variable is not correct. There are four cells in the table; but there are only two values for each of the two variables. Therefore there are two dichotomized variables (i.e., place of residence as one variable with rural and urban as the two categories; the second variable is diagnosis with psychoneurosis, depression, or schizophrenia as one category and mental retardation as the other category). Such a contingency table is called a 2 × 2 ("2 by 2") table and represents two dichotomized variables.

To compute chi-square, a null and an alternative hypothesis will have to be established, and the *expected* frequencies for each cell in the table must be calculated. (*Go to* 114)

---

113 There are two dichotomized variables in this table; two for place of residence and two for diagnosis. Such a contingency table is called a 2 × 2 ("2 by 2") table. To compute chi-square, a null and an alternative hypothesis will have to be established, and the *expected* frequencies for each cell in the table must be calculated. (*Go to* 114)

---

114 The intent of the chi-square statistic is to determine if the observed (empirically gathered) frequencies would have taken on the pattern that they do by chance or if there is some relationship between the two measured variables. Therefore the chi-square test has the ability to determine whether the two variables being measured are independent of one another.

The null and alternative hypotheses for the chi-square statistic follow the same logic as is followed in all other tests of statistical significance. Therefore the hypothesis statements would be:

H_0: Residence and diagnosis are independent of each other; that is, there are no differences among categories of variable A (here, place of residence) when compared across categories of variable B (here, psychiatric diagnosis).

H_A: Residence and diagnosis are not independent of one another.

It is not possible to specify the direction of the results in the chi-square statistic because even in the most simple chi-square table (2 × 2), there are four potential differences.

Once the null and alternative hypotheses have been established, you can proceed to calculate the *expected* frequencies. Essentially, the null hypothesis states that there is no relationship between the two variables. The purpose of the expected frequencies is to establish the number of cases that would have occurred in each of the cells if in fact the only fixed frequencies were the *marginals* (totals). That is, if the row totals (diagnosis) and column totals (residence) are accepted as given (89 rural and 272 urban residents; 184 psychoneurosis, depression, or schizophrenia, and 177 mental retardation diagnoses) and if the null hypothesis states that the two variables are independent, what frequencies would you expect within each cell? The chi-square statistic then compares these theoretical frequencies (expected frequencies) with the real frequencies (observed frequencies) to determine whether the size of the differences between the two could have occurred by chance.

The letters below and in the cells of Table 10-2 represent the observed frequencies for each cell in the 2 × 2 table and simultaneously give each cell a name. The expected frequency for any cell can be computed by multiplying the two marginals corresponding to the cell in question and dividing by N. Thus the expected frequency for cell (a) would be

$$(a + b)(a + c)/N = (184)(89)/361 = 16{,}376/361 = 45.4$$

or 45.4 is the expected frequency, by chance, of finding rural residents who are also psychoneurotic, depressive, or schizophrenic.

Copy Table 10-2 on a separate sheet of paper and try to compute the expected frequencies for the remaining cells ([b], [c], and [d]). Place the calculations below the obtained frequencies in parentheses. Be sure *not* to round off to whole numbers, but keep one decimal place. When you are finished . . .

(*Go to* 115)

---

115 The calculations for the expected frequencies of each cell are as follows. (Remember, each expected frequency is computed by multiplying the two marginals corresponding to the cell in question and dividing by N.)

$$\text{Cell (a)} = (a + b)(a + c)/N = (184)(89)/361 = 45.4$$
$$\text{Cell (b)} = (a + b)(b + d)/N = (184)(272)/361 = 138.6$$
$$\text{Cell (c)} = (c + d)(a + c)/N = (177)(89)/361 = 43.6$$
$$\text{Cell (d)} = (c + d)(b + d)/N = (177)(272)/361 = 133.4$$

Each of the expected frequencies should be placed in parentheses below the observed frequencies, which should produce a table similar to Table 10-3. You are now ready to calculate the chi-square test statistic.

If you want more practice determining expected frequencies . . .

(*Go to* 116)

If you want to calculate chi-square . . .

(*Go to* 118)

TABLE 10-3. Psychiatric diagnoses in Lucknow, India, by place of residence: obtained and (expected) frequencies*

	Place of residence		
Diagnosis	*Rural*	*Urban*	*Total*
Psychoneurosis, depression, schizophrenia	(a) 21 (45.4)	(b) 163 (138.6)	184
Mental retardation	(c) 68 (43.6)	(d) 109 (133.4)	177
Total	89	272	361

*Modified from Sethi, B. B., Nathawat, S. S., and Gupta, S. C. 1973. Depression in India. J. Soc. Psych. **91**:3-13 (Oct.).

---

116 What would be the expected frequencies for the first 2 × 4 table (Table 10-1) that was introduced? After calculating the expected frequencies, you can find the answers if you . . .

(*Go to* 117)

---

117 Your answers should be as follows:

Cell (a) = (a + b)(a + c + e + g)/N = (126)(89)/361 = 31.1
Cell (b) = (a + b)(b + d + f + h)/N = (126)(272)/361 = 94.9
Cell (c) = (c + d)(a + c + e + g)/N = (40)(89)/361 = 9.9
Cell (d) = (c + d)(b + d + f + h)/N = (40)(272)/361 = 30.1
Cell (e) = (e + f)(a + c + e + g)/N = (18)(89)/361 = 4.4
Cell (f) = (e + f)(b + d + f + h)/N = (18)(272)/361 = 13.6
Cell (g) = (g + h)(a + c + e + g)/N = (177)(89)/361 = 43.6
Cell (h) = (g + h)(b + d + f + h)/N = (177)(272)/361 = 133.4

(*Go to* 118)

118 The formula for chi-square essentially compares the differences between the observed frequencies and the expected frequencies relative to the expected frequencies, or:

$$\chi^2 = \Sigma \frac{(\text{Observed} - \text{Expected})^2}{\text{Expected}}$$

However, this formula can be greatly simplified for calculation purposes.
The calculation formula for chi-square is

$$\chi^2 = \Sigma \frac{f_o^2}{f_e} - N$$

where

χ^2 = chi-square
f_o = obtained frequency for each cell
f_e = expected frequency for each cell
N = total number of observations

To compute the chi-square for the dichotomized example in Table 10-3, the $\frac{f_o^2}{f_e}$ factor will have to be calculated for each cell (i.e., cell [a], [b], [c], [d]) and totaled before subtracting the N. Table 10-4 illustrates the calculation of the $\Sigma\frac{f_o^2}{f_e}$ term from the Indian mental health data. (*Go to* 119)

TABLE 10-4. Calculation table for $\Sigma \frac{f_o^2}{f_e}$, using Indian mental health data*

Cell	f_o		f_e	f_o^2	f_o^2/f_e
a	21		45.4	441	9.71
b	163		138.6	26,569	191.70
c	68		43.6	4,624	106.06
d	109		133.4	11,881	89.06
Total (N)	361	=	361.0	—	396.53

*Modified from Sethi, B. B., Nathawat, S. S., and Gupta, S. C. 1973. Depression in India. J. Soc. Psych. **91**:3-13 (Oct.).

119 Chi-square is easily obtained by subtracting N from the $\Sigma \frac{f_o^2}{f_e}$ factor. Thus:

$$\chi^2 = \Sigma \frac{f_o^2}{f_e} - N = 396.53 - 361$$

$$\chi^2 = 35.53$$

What does a chi-square value of 35.53 tell the researcher, and how does it help him make a decision about the null hypothesis?

The answer is, again, found by using a table of values, representing a sampling distribution for chi-square. The table used for the chi-square is slightly different. A chi-square table is included as Table 4 of Appendix A. Notice that in Table 4, the first column is the df column as in the t test table. Therefore before the results can be interpreted by using the table, the degrees of freedom must be determined. In previous cases the df was dependent on:

A. The number of cases sampled
B. The number of variables

(A, *go to* 120)
(B, *go to* 121)

---

120 The degrees of freedom for the t test was determined by the number of cases sampled. For contingency problems such as those encountered in chi-square, the degrees of freedom depend on the number of *cells* in the table. If r represents the number of rows and c represents the number of columns in the contingency table, the formula for finding the df is:

$$df = (r - 1)(c - 1)$$

Therefore in the example of a 2 × 2 table, the df would be 1 (i.e., df = [2 − 1] [2 − 1] = 1). What would be the df for a 3 × 3 table? When you have arrived at an answer . . .

(*Go to* 122)

---

121 The number of variables used in a t test had nothing to do with the degrees of freedom. In t tests you were always concerned with one variable, and that was the variable for which you calculated your $\overline{X}$ or $\overline{X}$'s in a two sample case. If you do not recall how this was done, it would be a good idea to review the t test segment again after finishing the program.

(*Go to* 120)

122 The df for the 3 × 3 table is 4. There are three columns and three rows:

$$\begin{aligned} df &= (r - 1)(c - 1) \\ &= (3 - 1)(3 - 1) \\ &= 4 \end{aligned}$$

The meaning of the concept of degrees of freedom can be easily seen in contingency problems like the chi-square. Essentially what it means is that once you have determined the value that falls in one cell (for a 2 × 2 table, df = 1) and given the marginals, all other cells will have fixed or predetermined values.

In Table 10-5, we can place any value between 0 and 10 in cell (a). Subsequent to that, all other cell values will be fixed. Hence in a 2 × 2 table, you have one cell that is free to vary, or you have one degree of freedom.

(*Go to* 123)

TABLE 10-5

(a)	(b)	10
(c)	(d)	10
10	10	20

123 Getting back to the decision about the chi-square of 35.53 for the Indian data, it can now be seen that if a significance level of .001 were set prior to the experiment and if there were a df of 1, the null hypothesis could be rejected. This is because:

A. The obtained value of chi-square is unlikely to occur by chance even beyond the .001 level of significance
B. The value 10.827 for df = 1, and α = .001 on the chi-square table is a likely outcome according to probability theory

(A, *go to* 124)
(B, *go to* 125)

124 Since the chi-square was 35.53, the result of the test statistic exceeds the needed 10.827 and is therefore significant at the .001 level.

It should be pointed out here that although a relationship is shown to exist between groups, this test statistic is insensitive to the direction of the relationship (i.e., the test involves the squares of deviations and cannot be negative).

Based on this finding, what conclusion would you draw? After deciding on a conclusion . . .

(*Go to* 126)

---

125 You may have misread the statement. It said that with a chi-square of 10.827, the outcomes were *likely*. On the contrary, if a chi-square of 10.827 or *greater* were obtained, there is little chance that no relationship exists between groups (i.e., H_0: $\chi^2 = 0$). Since the chi-square was 35.53, the result of the test statistic well exceeds the needed 10.827 and is significant at and beyond the .001 level.

It should be pointed out here that although a relationship is shown to exist, this test statistic is insensitive to the direction of the relationship (i.e., the test involves the squares of deviations and cannot be negative).

Based on this finding, what conclusion would you draw? After deciding on a conclusion . . .

(*Go to* 126)

---

126 You are able to reject the null hypothesis and conclude that with a known risk of committing a Type I (α) error ($p < .001$), the two variables, place of residence and type of diagnosis, are *not* independent of one another. In other words, there is some relationship between the two variables. If we examine Table 10-3, it is clear that diagnoses of mental retardation were relatively more prevalent in rural areas (68 out of 89 or 76%), whereas in urban areas only 109 out of 272 (or 40%) of those having a diagnosis were in the category of mental retardation.

Consider now the following hypothesis of interest:

H_A: An individual's perception of his/her health (measured as "good" or "poor") is related to a physician's perception of that person's health (also measured as "good" or "poor")

Lay out or construct the contingency table that you would use to display your frequency data. Note that each patient must fit into one and only one cell. Your contingency table should be similar to Table 10-6.

When you have finished . . .

(*Go to* 127)

---

127 In a study of persons on Old Age Assistance, Stahl (1971) took a random sample of O.A.A. recipients in Cook County, Illinois. After a physical examination by a team of physicians, nurses, and psychologists,

Continued.

each subject was given a "physician-perceived" health rating of "good" or "poor." Subsequently, without feedback to the subject, each was asked to rate his or her own health as "good" or "poor." At the end of the study, the data presented in Table 10-6 were obtained.

State the null and alternative hypotheses for this problem. When you are done . . .

(*Go to* 128)

TABLE 10-6. Self-perceived and physician-perceived health for sample of O.A.A. recipients*

Self-perceived	*Physician-perceived* *Good*	*Poor*	*Total*
Good	202	146	348
Poor	180	220	400
Total	382	366	748

*Modified from Stahl, S. M. 1971. Illness among the aged: a study of the determinants of the perception of levels of health in an indigent, urban population. Ph.D. dissertation. Urbana, Ill., University of Illinois.

---

128 H_0: There is no relationship between the patient's perception of his/her health and that of the physician

H_A: There is a relationship between the two perceptions of the patient's health at $\alpha = .05$

You should notice that the H_A statement does not make any claim to the direction of the findings. In other words, you are not saying that patients with "good" self-perceived health will be perceived by the physician as being in good health. To actually determine the "direction" of the findings, if they prove to be significant, you should always calculate percentages.

Given the hypothesis statements, calculate and fill in the expected frequencies for this research. When you have finished . . .

(*Go to* 129)

---

129 Your results should be as follows:

$$f_{e_a} = (348)(382)/748 = 177.7$$
$$f_{e_b} = (348)(366)/748 = 170.3$$
$$f_{e_c} = (400)(382)/748 = 204.3$$
$$f_{e_d} = (400)(366)/748 = \underline{195.7}$$
$$\Sigma f_e = \Sigma f_o = N = 748.0$$

Now calculate χ^2 and make a decision at $\alpha = .05$; then . . .

(*Go to* 130)

130 Your results should have been as follows:

Cell	f_o		f_e	f_o^2	f_o^2/f_e
a	202		177.7	40,804	229.62
b	146		170.3	21,316	125.17
c	180		204.3	32,400	158.59
d	220		195.7	48,400	247.32
	748	=	748.0		$\Sigma \frac{f_o^2}{f_e} = 760.70$

$$\chi^2 = \Sigma \frac{f_o^2}{f_e} - N = 760.70 - 748 = 12.7$$

$$df = (r - 1)(c - 1) = (2 - 1)(2 - 1) = 1$$

The chi-square table (Table 4) indicates that the calculated chi-square value of 12.7 is large enough to occur less than 5% of the time ($p < .05$). Therefore it is appropriate to reject the null hypothesis of no relationship between the two variables; the physician's perception of the patient's health is related to the patient's self-perception of his/her health.

Look again at Table 10-6 and determine the nature and direction of the relationship between the two assessments of the patient's health. When you have determined this . . .

(Go to 131)

131 It is apparent from Table 10-7 containing percentages for the two variables that those persons whose physical examination resulted in an assessment of good health also perceived themselves as being in good health. Similarly, those persons who were in poor health, as ascertained by the physician, considered themselves more frequently in the poor health category by their own assessment. The two perceptions are not independent of one another, and the convergence between the two variables is statistically significant.

(Go to 132)

TABLE 10-7. Self-perceived and physician-perceived health for sample of O.A.A. recipients, percentages*

	Physician-perceived			
	Good		*Poor*	
Self-perceived	N	%	N	%
Good	202	52.9	146	39.9
Poor	180	47.1	220	60.1
Total	382	100.0	366	100.0

*Modified from Stahl, S. M. 1971. Illness among the aged: a study of the determinants of the perception of levels of health in an indigent, urban population. Ph.D. dissertation. Urbana, Ill., University of Illinois.

132 There are two cautions to introduce about the chi-square statistic beyond the fact that it does not indicate direction nor degree of relationship. One is that although the significance level of chi-square is based only on the df (number of cells), the statistic is not independent of sample size. Consider a contingency table in which one of the cells has an observed frequency of 15 and an expected value of 10. $\chi^2 = \frac{(15 - 10)^2}{10} = 2.5$. Now if the number of cases is increased 10 times throughout the table, then the observed frequency becomes 150 and the expected value becomes 100. Now $\chi^2 = \frac{(150 - 100)^2}{100} = \frac{2500}{100} = 25$.

You are more likely to obtain statistically significant values with large samples.

The second caution to mention before concluding the chi-square section is that the test assumes independence between columns and between rows. That is, a subject in row 1 should not also be in row 2. A typical case of nonindependence occurs when you take repeated measures from the same subject (e.g., in a before-after experimental situation).

You have now completed the substantive section of the chapter on specific tests of statistical significance. Before going on, it is appropriate to present a few ideas on how to distinguish between the tests presented in this chapter when you are faced with research problems that require a choice of tests or when you are reading reports of research and attempting to determine the appropriateness of the tests used. These few remaining frames attempt to present a few of the ground rules for helping you make this choice.

(*Go to* 133)

133 Rule 1. You must decide whether a specific problem calls for a measure of association (i.e., *how strongly* are two variables related?) or whether it calls for a test of statistical significance (i.e., *is there a relationship* or difference between two variables or between a statistic and a parameter?). Association will be covered in the next chapter, so this choice will be more fully explained later.

(*Go to* 134)

134 Rule 2. If you have decided on using a test of statistical significance, the next decision is whether you are dealing with two separate variables. If you are interested in the existence of a significant relationship between race and a given test score, for example, you are dealing with two different variables. If you are interested in the existence of a significant relationship between the mean height of a sample and the population's mean height, you are dealing with only one variable. In the two-variable case you should use the chi-square statistic; in the one-variable case you should use some form of the Z or t statistic.

(*Go to* 135)

---

135 Rule 3. Another basis for making a decision between chi-square and Z or t is the level of measurement used for measuring that variable(s). The Z and t require at least an interval level of measurement so that the mean ($\overline{X}$) for that variable can be calculated. The chi-square requires only a nominal level of measurement. Associated with the level of measurement is the researcher's ability to use means or proportions. The Z test requires large sample sizes for its use ($N > 30$), and only when the N is greater than approximately 50 should proportions be calculated. Therefore, with large N's, nominal data, and one variable, it is possible to calculate proportions and therefore use the Z test.

(*Go to* 136)

---

136 Rule 4. Perhaps the simplest rule to remember and understand is that when dealing with interval levels of measurement ($\overline{X}$'s), you have a choice between Z and a t test. With large N's ($N > 30$), use the Z test, and with small N's ($N \leq 30$), use the t test. The central limit theorem and the more conservative nature of the t statistic provide an explanation for this distinction.

(*Go to* 137)

---

137 Following are a few examples to help you in determining which statistic to use. The distinction between association and statistical significance is not critical here, so these examples all relate to the existence of statistical significance only.

A drug firm reports the reduction of diastolic blood pressure as a mean of 17 points with a standard deviation of 3 points based on their clinically controlled trials of 143 patients. Your trials in the outpatient department ($N = 27$) indicate that there is a reduction of 13 points with a standard deviation of 5 points. Which test would you use to indicate whether there is a difference between the two groups?

A. t test for differences of means
B. Z test for differences of means
C. t test for sample vs. population mean
D. Chi-square

(A, *go to* 138) (C, *go to* 140)
(B, *go to* 139) (D, *go to* 141)

---

138 1. You were dealing with mean diastolic blood pressure, so you could eliminate chi-square, since you have a higher level of measurement than the nominal level generally used with chi-square.

2. You were dealing with two samples, so you should eliminate the sample vs. population mean choice.

3. You were dealing with a large sample (N = 143) and a small sample (N = 27), and therefore you should choose the more conservative test, the t test for differences between means, rather than the Z test of differences between means, since the latter requires that both of the samples have large sample sizes (N > 30).

The t test for differences of means is correct.

(*Go to* 142)

---

139 You were dealing with a large sample (N = 143) and a small sample (N = 27), and therefore you should choose the more conservative test, the t test for differences between means, rather than the Z test of differences between means, since the latter requires that both of the samples have large sample sizes (N > 30).

(*Go to* 138)

---

140 You were dealing with two samples, so you should eliminate the sample vs. population mean choice.

(*Go to* 138)

---

141 You were dealing with the mean diastolic blood pressure, so you could eliminate chi-square, since you have a higher level of measurement than the nominal level generally used with chi-square.

(*Go to* 138)

---

142 Is there a significant difference between the finding that 11% of a sample (N = 849) of a city are black and the census report indicating that the proportion of blacks in the city is 0.09?

What test of statistical significance would you use?

A. Z test for differences of proportions
B. t test for differences of proportions
C. Chi-square
D. Z test for sample vs. population proportions

(A, *go to* 143) (C, *go to* 145)
(B, *go to* 144) (D, *go to* 146)

143 Z tests for differences of proportions should be eliminated because this test assumes that you have two independent samples; you are dealing with a sample and a population.

(*Go to* 146)

144 The t test for the difference of proportions should have been eliminated because this test assumes that you have two independent samples; you are dealing with a sample and a population. Furthermore, you are dealing with a large sample size (N = 849), and you can make use of the central limit theorem in selecting a test of statistical significance. The t test does not directly allow for this.

(*Go to* 146)

145 The chi-square test should have been eliminated because you are dealing with only one variable—proportion (percent) black. The chi-square test requires that two variables be used.

(*Go to* 146)

146 The correct answer is the Z test for sample vs. population proportions. The chi-square test was eliminated because you are dealing with only one variable—proportion (percent) black. The Z test for differences of proportions was eliminated because this test assumes that you have two independent samples; you are dealing with a sample and a population. The t test for differences of proportions was eliminated for much the same reason that the Z test for differences of proportions was eliminated plus the fact that the sample size was large (N = 849).

(*Go to* 147)

147 In a given school system, is there a significant relationship between IQ and race? To answer this question you have selected a sample of 738 fourth- and fifth-graders and measured their IQ with a standard test procedure.

Which test statistic would you use?

A. Z test for difference of means
B. t test for difference of means
C. Z test for sample vs. population means
D. Chi-square

(A, *go to* 148) (C, *go to* 150)
(B, *go to* 149) (D, *go to* 151)

148 You could use the Z test for difference of means by comparing mean IQ's for two races (e.g., black vs. white). But perhaps the more appropriate test statistic would have been chi-square, since you are dealing with two variables, one of them nominal (race) and the other interval (IQ). If more than two races are found in the school, then the Z test for difference of means cannot be used.

(*Go to* 152)

149 The t test for difference of means could have been used by comparing mean IQ's for two races (e.g., black vs. white). But perhaps the more appropriate test statistic would have been chi-square, since you are dealing with two variables, one of them nominal (race) and the other interval (IQ). If more than two races are found in the school, then the t test for difference of means cannot be used. Furthermore, if each category of race had more than 30 students in it, the Z test would have been more appropriate than the t test; a subcategory of race with an $N \leq 30$ would have suggested the t test.

(*Go to* 152)

150 There is no sample vs. population situation in this problem. In reality you could use the chi-square test, since you are dealing with two variables, one of them being nominal (race), the other interval (IQ), or you could have used the Z test or the t test for differences of means, with the means being IQ for each racial group. If each category of race had more than 30 students in it, the Z test would have been appropriate; a subcategory of race with an $N \leq 30$ would have suggested the t test for the differences of means.

(*Go to* 152)

151 In reality you could use the chi-square test, since you are dealing with two variables, one of them being nominal (race), the other interval (IQ), or you could have used the Z test or the t test for differences of means, with the means being IQ for each racial group. If each category of race had more than 30 subjects in it, the Z test would have been appropriate; a subcategory of race with an $N \leq 30$ would have suggested the t test for the differences of means.

(*Go to* 152)

152 You are interested in finding whether the type of organizational leadership (authoritarian vs. democratic vs. laissez-faire) has an effect on or is related to the perception of employees regarding the capability of leaders. Employees are asked whether they feel that their immediate supervisors are effective or ineffective leaders. Your sample consists of 67 persons distributed among three types of organizational leadership. Which statistical test would you use?

A. t test for the difference of proportions
B. Chi-square
C. Z test for the difference of proportions
D. Z test for the difference of means

(A, *go to* 153) (C, *go to* 154)
(B, *go to* 156) (D, *go to* 155)

---

153 There is no t test for the difference of proportions. The calculation of proportions depends on sample sizes (or populations) that are fairly large, and the t test uses small sample sizes.

(*Go to* 156)

---

154 The Z test for the difference of proportions is not appropriate, since you are dealing with two variables.

(*Go to* 156)

---

155 The nominal nature of the two variables indicates that means cannot be calculated. The Z test for the difference of means is therefore inappropriate.

(*Go to* 156)

---

156 The only appropriate test is chi-square. You are dealing with a comparison of two nominal variables (type of organizational leadership and perceived effectiveness). The nominal nature of the variables indicates that means could not be calculated, so the Z test for difference of means is inappropriate. The Z test for difference of proportions is not appropriate, since you are dealing with two variables, and there is no such test as the t test for the difference of proportions.

(*End of chapter* 10)

---

chapter
11

Association

At the conclusion of this chapter, you should be able to:

1. Distinguish between correlation problems and prediction problems within association by correctly indicating whether a given problem or hypothesis calls for a correlation or a prediction procedure.
2. State in which of two variable data sets a measure of association would be appropriate (bivariate distributions or logically paired data) and in which sets a measure of association would not be appropriate (univariate distributions or independent data sets).
3. Given examples of bivariate data sets, determine whether a product-moment correlation or a rank-order correlation would be appropriate, based on the scale of measurement of the variables.
4. Graphically represent the relationship between two sets of scores by constructing a scatter diagram.
5. Discuss the following with respect to the product-moment correlation coefficient (r):
   a. The distinction between correlation and causation
   b. The distinction between linear and nonlinear association
   c. The distinction between positive and negative correlation
   d. The distinction among zero, moderate, and perfect correlations
   e. The numerical values of a perfect correlation
   f. The independence of the coefficient from sample size, variable mean, and units of measurement
   g. The scale of measurement of the variables on which r can be applied
6. Estimate the direction and magnitude (within $\pm.3$) of r from a scatter diagram.
7. Read and interpret a matrix of intercorrelations in terms of direction and relative magnitude of the coefficients.
8. Compute a correlation coefficient (r) given raw data on two variables from the same sample.
9. Do the following with respect to r:
   a. State the usual null hypothesis for r (H_0: $r = 0$).

b. Use a table giving df, α, and r to determine the significance of a given r.
c. Construct an illustrative hypothesis requiring the use of a one-tailed test of significance for r.

10. Calculate the 95% confidence interval for r, given r, N, and a table for the r-to-Z transformations.
11. Write the general formula for a straight line ($Y = a + bX$) and define the slope (b, an indicator of steepness and direction) and the Y intercept (a, the point on Y where the line crosses Y).
12. Plot the graph of a given straight-line (regression) equation, and calculate a predicted Y score, given an X score.
13. State which variable is the dependent or criterion variable and which variable is the independent or predictor variable, when given an English statement concerning predicting variable A from information on variable B.
14. Determine from a scatter plot whether or not a linear regression line is appropriate.
15. Compute the equation for the regression of X on Y for a given two-variable set of data.
16. Assign rank-orders to a set of scores including tied scores.
17. Calculate a rank-order correlation coefficient (rho or ρ) for a given set of data.
18. Determine the significance of a given rank-order correlation coefficient, using a table giving N, α, and ρ.
19. Link from memory the following symbols with a list of terms: p, $s_{\bar{X}}$, H_0, α, Z, t, χ^2, df, r, and ρ.

---

1 Measures of association answer the question: Within a given population, is there a *relationship* between one variable (or set of variables) and another variable (or set of variables)? The measure of association is generally performed on sample data from the population and actually is a descriptive statistic; that is, it describes something about the sample. Measures of association are frequently tested for statistical significance to determine the probability of that particular result happening by chance alone. Therefore in this chapter both descriptive statistics (measures of association) and inferential statistics (the significance of these measures of association) will be presented.

A measure of association can exist only if data can be logically paired in some way. That is, for each measure on variable X we must be able to associate with it one and only one measure on variable Y. Such data are usually said to have a bivariate distribution. Bivariate data may arise

Continued.

from the same measure being made at two different times or places on the same population or from two different measures being made on the same population.

It can be seen that measures of association can occur only if we have data on two or more variables from the (same/different) population(s).

A. Different

B. Same

(A, *go to* 2)
(B, *go to* 4)

---

2 There is no way to logically pair data from two different populations. We cannot link a particular maze-running ability score from a sample of rats to a particular intelligence score from a sample of human patients.

Could we get a measure of association between college entrance exam scores of freshman men and freshman women at a given university?

A. Yes

B. No

(A, *go to* 3)
(B, *go to* 4)

---

3 There is no way to link the college entrance exam score for a particular woman with any particular male score. The data are not logically paired as is necessary before an association can be made. Association measures can be made when we have two or more clearly defined variables from the same population.

(*Go to* 4)

---

4 Association measures require two or more clearly defined variables from the *same* population (sample). Note, however, that a population may be complex such as a father-son pair (baldness in father related to baldness in son), or a time-space link (rain in 1974 for each county related to corn production in 1975 for the same counties).

If we can find that the smoking of cigarettes during an individual's early lifetime is highly associated with lung cancer at age 50, can we say this is proof that smoking causes lung cancer?

A. Yes

B. No

(A, *go to* 5)
(B, *go to* 6)

---

5 Association never proves causation. The fact that X is related to Y does not mean that X causes Y. Association measures are reciprocal: the same data and measures allow you to say both "hypertension is related to heart disease" and "heart disease is related to hypertension." You *can-*

not say from the data which causes which. In statistical language, measures of association do not imply causality. It is only the theory, other research findings, or the time sequence that allows us to say that smoking "causes" lung cancer.

(*Go to* 7)

---

6 Measures of association do not imply causality. Causation must be established by other logical analyses, not from a measure of association. All we can say from a measure of association is that smoking and lung cancer are associated variables.

(*Go to* 7)

---

7 In addition to distinguishing between association and causation, we must distinguish between correlation and prediction. Both correlation and prediction fall within association.

Correlation answers the question, "What is the degree of relationship between variable X and variable Y in a given population?"

Prediction answers the question, "If I know score X for a person, how accurately can I predict score Y?"

The two questions are highly related, but they call for somewhat different analyses. Is the problem stated below a correlation problem or a prediction problem?

> A psychologist wants to use data from a large sample of mental patients to provide the basis whereby in the future he can estimate the duration of hospitalization for a patient by knowing his or her severity of diagnosis score and social class.

A. Correlation problem
B. Prediction problem

(A, *go to* 8)
(B, *go to* 10)

---

8 The psychologist's problem is not really a correlation problem. He does not want to know the degree to which the variables are associated (co-related). Although that information may be useful, the task called for is to estimate some characteristic about an individual. Estimation, in this case, is prediction. Given severity of diagnosis and social class, he wants to predict duration of hospitalization.

If the question raised was "Is there a relationship between severity of diagnosis and social class, and duration of hospitalization in my sample of patients?", then this would be a *correlation/prediction* problem.

Make your decision between correlation and prediction and then . . .

(*Go to* 9)

---

9 The answer to "Is there a relationship?" is in terms of the degree of that relationship. The degree of relationship, which is a correlation problem, does not tell us how to predict a particular score. For that we need other information, which will be presented later.

(*Go to* 11)

10 Given data for a particular variable, you want to predict or estimate the value of another variable; the psychologist's problem is a prediction situation. The correlation problem might be: "Is there a relationship between duration of hospitalization, and severity of diagnosis and social class in my sample of patients?"

(*Go to* 11)

11 There are many different specific measures of association. Which one to use depends on (1) whether it is a prediction problem or a correlation problem, (2) whether there are two or more than two variables, and (3) what the scales of measurement of the variables are.

The following paradigm presents some of the measures in the various categories.

		Variable 1		
		Interval	*Ordinal*	*Nominal*
	Interval	r		
Variable 2	*Ordinal*		ρ, γ, τ	
	Nominal	r_b, r_{pb}		C, λ, ϕ

where

r (product-moment correlation) is the most common correlation measure
$\mathbf{r_b}$(biserial r) and $\mathbf{r_{pb}}$ (point biserial r) are variations used when one of the variables is a nominal scale
ρ (rho), γ (gamma), and τ (tau) are three measures of association usable on ordinal data
C (contingency), λ (lambda), and ϕ (phi) may be used to correlate two nominal scales

All of these have certain restrictions in their uses.

This book will discuss only two of the most frequently used correlation measures and one prediction method: r, the product-moment correlation between two interval scale variables; ρ (rho), the rank-order correlation between two ordinal scale variables; and *regression analysis*, a method that uses information on interval scale variables X and Y from a sample to allow you to predict Y for any given X for a person in the population.

If you wish to start the frames on r . . .

(*Go to* 12)

If you wish to skip r and go to regression analysis . . .

(*Go to* 58)

If you wish to skip both the r and regression analysis frames and go to ρ . . .

(*Go to* 81)

---

12 This frame begins the section on the *product-moment correlation*, or r. Consider the following array of data:

Subject	X	Y
A	30	22
B	50	38
C	10	8
D	60	45

Each person's scores can be plotted on a two-dimensional graph such as found in Fig. 11-1. Note that each point on the graph represents a score on X and a score on Y. Here the points seem to lie on a ________ line.

Make a decision on how you would describe that line, and then . . .

(*Go to* 13)

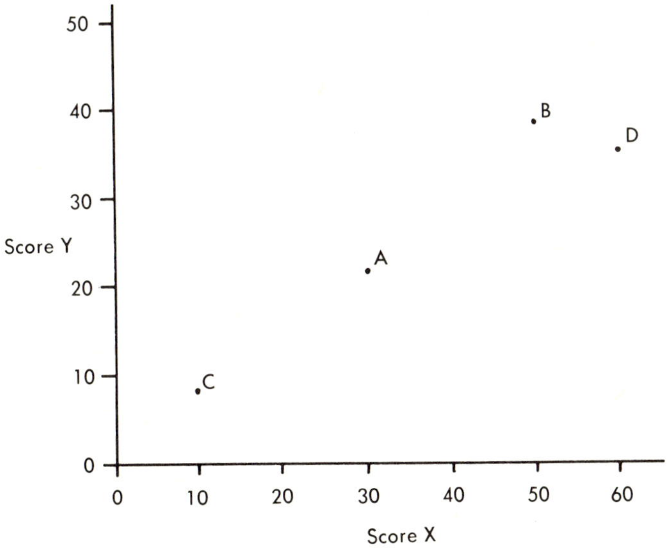

Fig. 11-1. Scatter diagram I.

13 The best way to describe Fig. 11-1 would be that the points approximate a straight line. This section will be dealing with information on two variables that can be plotted as a straight line (linear) relationship. Curvilinear relationships will not be discussed. The two-variable graph of data is called a *scatter diagram.* If we wish to add a fifth person to the scatter diagram in Fig. 11-2 with an X score of 40 and a Y score of 30, at which point should that person be placed?

A. Position 3
B. Position 1
C. Position 2

(A, *go to* 14)
(B, *go to* 16)
(C, *go to* 14)

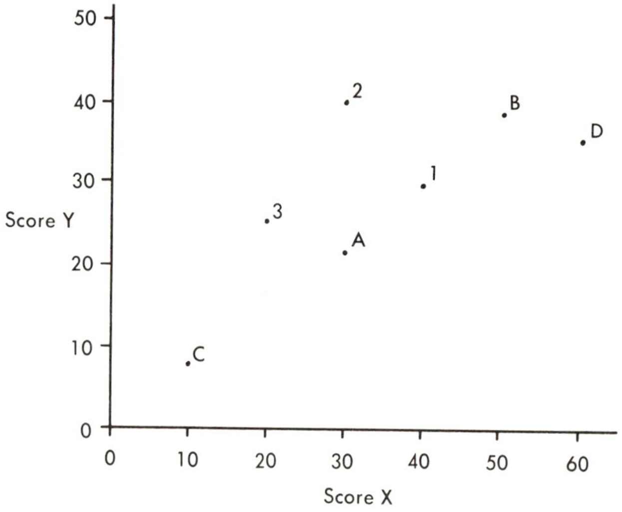

Fig. 11-2. Scatter diagram II.

14 Look carefully at the scatter diagram. You wanted the position for X = 40 and Y = 30, and your decision was wrong. Position 2 represents X = 30, Y = 40. Position 1 represents X = 40, Y = 30, which is the location we want.

What scores does position 3 represent?

X = _____________, Y = _____________

Make your decision and then . . .

(*Go to* 15)

---

15 Position 3 represents X = 20, Y = 25.

(*Go to* 17)

---

16 Position 1 represents X = 40, Y = 30; your answer is correct.

(*Go to* 17)

---

17 A scatter diagram shows the joint distribution of two variables. It helps us visualize the two concepts in a correlation coefficient: *direction* and *magnitude*.

With regard to direction, when an increase in the measurement of one variable is accompanied by an increase in the measurement of another variable, the two variables are correlated ________________.

A. Negatively
B. Positively

(A, *go to* 18)
(B, *go to* 19)

---

18 Two variables are correlated *positively* when an increase in one is associated with an increase in the other, or when a decrease in one variable is accompanied by a decrease in the other variable. As level of educational attainment increases, income also tends to increase.

Two variables are correlated *negatively* when an increase in one variable is accompanied by a decrease in the other variable. An example of a negative correlation is the situation in which increased income is found to be correlated with decreased death rates.

(*Go to* 20)

---

19 If two variables change in the same direction, their correlation is *positive*. For example, incomes increase as education increases. If they change in opposite directions (that is, if the two variables are inversely related), their correlation is *negative*. Death rates are lower in higher income groups.

(*Go to* 20)

---

20 The second concept involved in correlation, in addition to direction, is the concept of *magnitude*. Magnitude in correlation refers to the *degree of relationship* between two variables.

The magnitude of a product-moment correlation coefficient ranges from −1.00 to +1.00. An r of +1.00 indicates a "perfect" positive correlation. That is, the variation in one variable is perfectly related to the variation in the other variable, or they are said to co-vary together. A scatter diagram of two variables with an r = +1.00 would be similar to that found in Fig. 11-3.

(*Go to* 21)

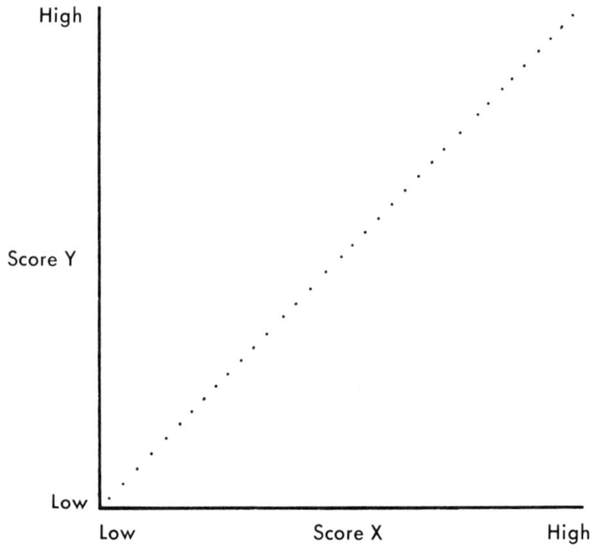

FIG. 11-3. Scatter diagram of perfect positive relationship (r = +1.00).

21 Which scatter diagram in Fig. 11-4 has an r = −1.00?

A. Scatter diagram A
B. Scatter diagram B
C. Scatter diagram C

(A, *go to* 22)
(B, *go to* 23)
(C, *go to* 24)

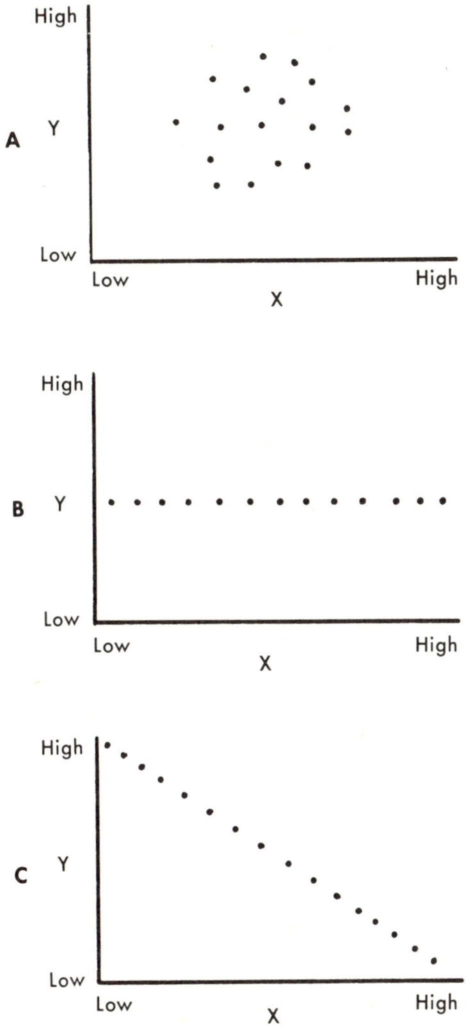

FIG. 11-4

22 Scatter diagram A does not show a correlation of −1.00. A −1.00 correlation has a magnitude of 1.00 (a perfect correlation) and a negative direction. An r of −1.00 occurs when each unit of increase in one variable is accompanied by an equivalent unit of decrease in the other variable. Scatter diagram C in Fig. 11-4 illustrates an r = −1.00.

(*Go to* 24)

23 Scatter diagram B is not the appropriate answer; it does not depict r = −1.00. A −1.00 correlation has a magnitude of 1.00 (a perfect correlation) and a negative direction. An r of −1.00 occurs when each unit of increase in one variable is accompanied by an equivalent unit of decrease in the other variable. Scatter diagram C in Fig. 11-4 illustrates an r = −1.00.

(*Go to* 24)

---

24 A perfect negative correlation (r = −1.00) is found in scatter diagram C of Fig. 11-4.

Two variables that are totally unrelated to each other are called "uncorrelated." Two uncorrelated variables will produce an r = 0.00. Which of the scatter diagrams in Fig. 11-5 represents r = 0.00?

A. Scatter diagram A
B. Scatter diagram B
C. Scatter diagram C

(A, *go to* 25)
(B, *go to* 26)
(C, *go to* 27)

---

25 Scatter diagram A in Fig. 11-5 shows absolutely no change in variable Y as variable X increases in magnitude. Therefore a zero correlation would result. The other manner of obtaining a zero correlation is seen in scatter diagram C of Fig. 11-5 where any possible change in variable Y might be found for a change in variable X. Only scatter diagram B in Fig. 11-5 depicts a correlation greater than zero, since an increase in variable X seems to be associated with an increase in variable Y.

(*Go to* 28)

---

26 Scatter diagram B in Fig. 11-5 depicts a correlation greater than zero, since an increase in variable X seems to be associated with an increase in variable Y.

There are two ways to obtain a zero correlation:

1. Scatter diagram A in Fig. 11-5 shows absolutely no change in variable Y as variable X increases in magnitude.
2. Scatter diagram C in Fig. 11-5 shows that any possible change in variable X can be associated with any possible value of variable Y.

(*Go to* 28)

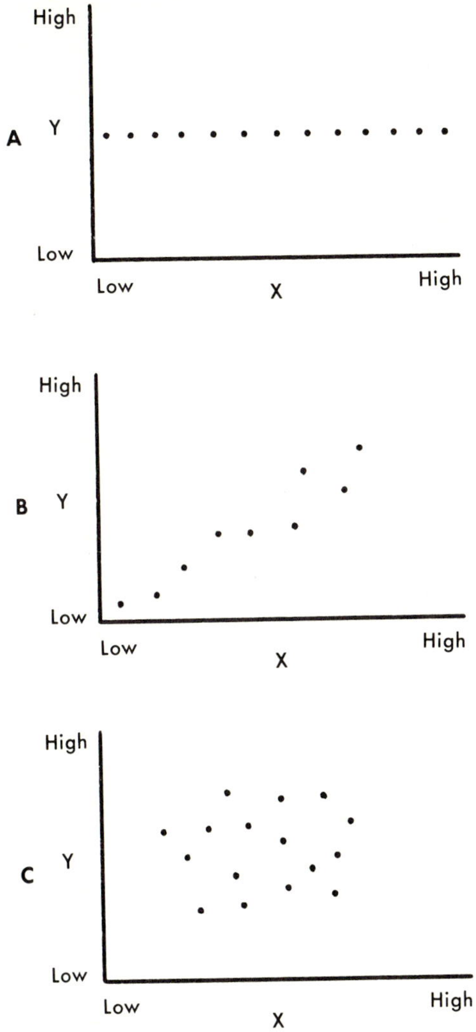

FIG. 11-5

---

27 Scatter diagram C in Fig. 11-5 depicts a zero correlation situation in which any possible change in variable Y might be found for a change in variable X. Scatter diagram A, however, also depicts a zero correlation situation in which there is absolutely no change in variable Y as variable X increases in magnitude. Only scatter diagram B in Fig. 11-5 depicts a correlation that is greater than zero, since an increase in variable X seems to be associated with an increase in variable Y.

(*Go to* 28)

---

28 The numerical values of correlation coefficients, in real data, fall between 0.00 and ±1.00. Coefficients in the r = ±.30 to r = ±.70 range could be called "moderate" correlations. When a correlation is not perfect, not all the points in a scatter diagram lie in a sloped straight line. The degree (magnitude) of a correlation reflects the relative tightness or nearness of the points around a straight line.

Fig. 11-6 depicts several scatter diagrams with estimates of the magnitude and direction of the correlation coefficient, r. Note that as the plotted data become more spread out, the correlation decreases.

(*Go to* 29)

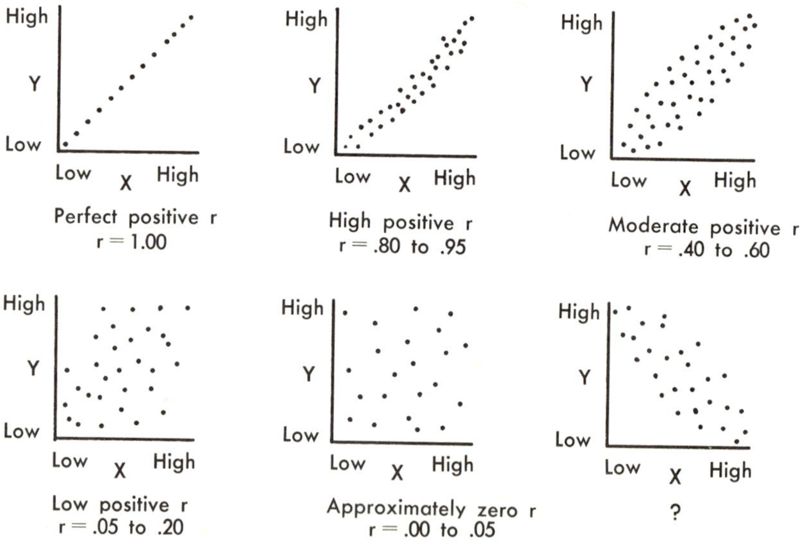

FIG. 11-6

29 Now look at the three scatter diagrams in Fig. 11-7 and estimate the magnitude of the correlation coefficient and the sign or direction of that coefficient. You should be able to estimate the value of r within ±.3. When you have completed that . . .

(*Go to* 30)

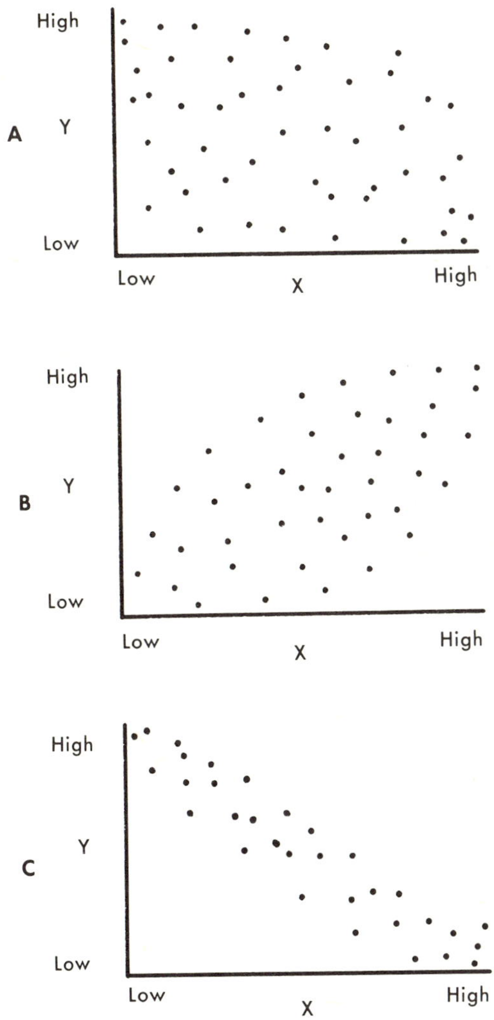

FIG. 11-7

30 We estimate that the three scatter diagrams in Fig. 11-7 approximate the following r values: A, $r = -.10$; B, $r = +.40$; C, $r = -.90$.

Next, let us consider what is usually called a matrix of intercorrelations. When several variables are used in a study, you may compute a correlation coefficient between all possible pairs of variables. The result is displayed in a grid or matrix such as the following:

		X	Y	Z
	X	1.00	.73	.25
Variable	Y	.73	1.00	−.61
	Z	.25	−.61	1.00

The correlation coefficient between variable X and variable Y (r_{xy}) is .73. Note that whether we speak of r_{xy} or r_{yx} the coefficient has the same value. Note also that the correlation of any variable with itself is a perfect correlation. Which statement below best describes the association between variable Y and variable Z?

A. High positive correlation
B. Moderate negative correlation
C. Low positive correlation

(A, *go to* 31)
(B, *go to* 32)
(C, *go to* 33)

---

31 The correlation between variables Y and Z is $r = -.61$. You should return to the previous frame where you should consider carefully: (1) the location on the matrix where r_{yz} can be found, (2) the direction of the correlation, and (3) the magnitude of the coefficient.

(*Go to* 30)

---

32 A correlation of $r = -.61$ is best described as a moderate negative correlation.

(*Go to* 34)

---

33 You have made an error in your decision; the correlation between variables Y and Z is $r = -.61$. You should return to frame 30 where you should consider carefully: (1) the location on the matrix where r_{yz} can be found, (2) the direction of the correlation, and (3) the magnitude of the coefficient.

(*Go to* 30)

---

34 Before turning to the computation of r, let us reconsider the question of *linearity* mentioned earlier. The product-moment correlation measures the degree of fit or closeness of the data to some best-fit imaginary straight line. (The best fit is based on the concept of "least-squares" and will be dealt with under the regression section later in this chapter.) Which scatter diagram in Fig. 11-8 has the higher product-moment correlation?

A. Scatter diagram A (A, *go to* 35)
B. Scatter diagram B (B, *go to* 36)

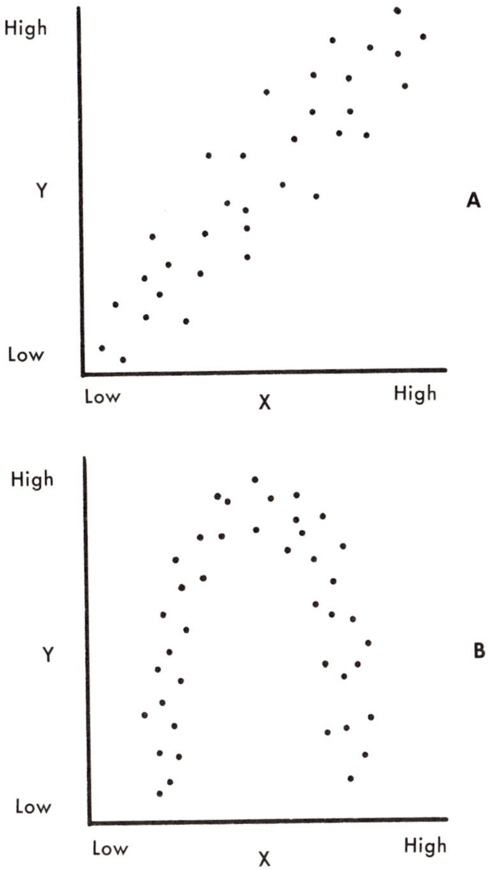

FIG. 11-8

---

35 Diagram A in Fig. 11-8 would have a higher product-moment correlation than diagram B. Diagram B displayed a strong *curvilinear* relationship between the two variables but a negligible linear relationship. The product-moment correlation deals with linearity; curvilinear techniques will not be discussed in this text. More advanced statistics texts deal with this issue.

(*Go to* 37)

---

36 Diagram B in Fig. 11-8 illustrates a strong curvilinear relationship. But the points are considerably scattered from any straight line that might be plotted through the data, and r measures the degree of adherence of the data to a straight line. The points in diagram A lie much closer to a straight line. Therefore diagram A will have a higher product-moment correlation coefficient.

(*Go to* 37)

---

37 Next we will discuss the calculation of the correlation coefficient, or r, where:

$$r = \frac{\Sigma(X - \overline{X})(Y - \overline{Y})}{N s_x s_y}$$

There is an easier computational formula for r, but first look at what r means. The term $(X - \overline{X})(Y - \overline{Y})$ is called a *cross product*. It is the product of the deviation of an X score from its mean and the deviation of a corresponding Y score from its mean. (Each of the components of the cross product should be familiar to you from your knowledge of the standard deviation.) As the deviations get larger, the cross product gets larger. A correlation is a way to look at the size of the cross product or the deviations of the scores about their respective means.

Any two variables, X and Y, have unique means and standard deviations. We want the correlation coefficient to measure association independently of the particular variables and particular sample; i.e., what is desired is a generic concept for measuring correlation.

To make our coefficient independent of sample size, variable mean, and unit of measurement, we take into account the value of N, $\overline{X}$, and s (standard deviation). In other words, the product-moment correlation coefficient is the sum of the cross products of the deviation scores corrected for sample size and units of measurement.

If the degree of association between age (variable A) and blood pressure (variable B) in a sample of 4000 patients is the same as the degree of association between income (variable I) and distance from place of employment (variable D) in a sample of 1672 respondents, which statement below is true?

A. $r_{AB} > r_{ID}$
B. Not enough information given (A, *go to* 39)
C. $r_{AB} = r_{ID}$ (B, *go to* 40)
(C, *go to* 38)

---

38 The same degree of association between the two sets of variables would result in the following:

$$r_{AB} = r_{ID}$$

A formula for r that is much easier to compute can be expressed in the original (raw) scores:

$$r = \frac{N\Sigma XY - \Sigma X \Sigma Y}{\sqrt{[N\Sigma X^2 - (\Sigma X)^2]\,[N\Sigma Y^2 - (\Sigma Y)^2]}}$$

To compute this formula you will need an X score and a Y score for each observation, as well as the products XY, X^2, and Y^2.

Table 11-1 should be completed so that you can calculate r. Complete the table by calculating the remaining X^2, Y^2, and XY values for each person. Then sum each column. The sums represent:

ΣX ΣX^2 ΣXY

ΣY ΣY^2

When your calculations are complete . . . (*Go to* 41)

TABLE 11-1. Self-image and reading achievement scores* from a small sample of first-grade pupils

Pupil	*Self-image* X	*Reading achievement* Y	X^2	Y^2	XY
1	8	8	64	64	64
2	2	3	4	9	6
3	10	9	100	81	90
4	7	7	49	49	49
5	5	4	25	16	20
6	6	6			
7	2	2			
8	10	9			
9	3	3			
10	8	9			
	Total $\Sigma X = 61$	$\Sigma Y =$	$\Sigma X^2 =$	$\Sigma Y^2 = 430$	$\Sigma XY =$

*Both self-image and reading achievement are intervally measured variables; both range between 0 and 10, with 10 being high.

39 The problem posed said that the degree of association was exactly the same in both cases. The units of measurement or the differences in sample sizes do not matter. Both r's would be the same.

(Go *to* 38)

40 Enough information was provided for you to make your decision. The problem posed stated that the degree of association was exactly the same in both cases. The units of measurement or the differences in sample sizes do not matter. Both r's would be the same.

(*Go to* 38)

41 The values you calculated should be:

$$\Sigma X = 61 \qquad \Sigma X^2 = 455 \qquad \Sigma XY = 440$$
$$\Sigma Y = 60 \qquad \Sigma Y^2 = 430$$

If your results do not agree, check your calculations. Then proceed to calculate the product-moment correlation coefficient:

$$r = \frac{N\Sigma XY - \Sigma X\Sigma Y}{\sqrt{[N\Sigma X^2 - (\Sigma X)^2][N\Sigma Y^2 - (\Sigma Y)^2]}}$$

When you are ready to check your calculations . . .

(*Go to* 42)

42

$$r = \frac{10 \cdot 440 - 61 \cdot 60}{\sqrt{[10 \cdot 455 - (61)^2][10 \cdot 430 - (60)^2]}} = \frac{4400 - 3660}{\sqrt{[4550 - 3721][4300 - 3600]}}$$
$$= \frac{740}{\sqrt{[829][700]}} = \frac{740}{\sqrt{580{,}300}} = \frac{740}{762} = .971$$

A correlation coefficient of r = .971 indicates that (1) the two variables are positively related; i.e., as self-image increases in value, so does the pupil's reading achievement score, at least for this sample; and (2) the two variables are strongly related; i.e., the clustering of the observations about a best-fit line would be tight, since the calculated r is close to a perfect correlation of 1.00.

Although the computation of r may take some time, the important tasks are in making sure that the computation was appropriate in the first place, and in interpreting it in the second place. Remember that the correlation question asks, "What is the degree of relationship between variable X and Y?" Our r value *describes* a relationship in our sample of data. Most often, however, we are more interested in making some *inference* about the population from which the sample was drawn.

Our inference may be a test of a hypothesis regarding the correlation coefficient, or it may be an estimate of the population r or interval about r.

The basic inferential test of r posits a null hypothesis that there is no correlation between variable X and variable Y. Which statement below expresses this null hypothesis?

A. H_0: $r < 0$
B. H_0: $r_x \neq r_y$
C. H_0: $r = 0$

(A, *go to* 43)
(B, *go to* 44)
(C, *go to* 45)

---

43 The statement $r < 0$ is a hypothesis that there is a negative correlation between variable X and variable Y. Your answer is not correct. A "no correlation" hypothesis is represented by $r = 0$.

(*Go to* 45)

---

44 The statement $r_x \neq r_y$ does not make sense. Since r relates two variables, you cannot have r_x or r_y but rather r_{xy}. A "no correlation" hypothesis is represented by $r = 0$.

(*Go to* 45)

---

45 The basic null hypothesis for testing a correlation coefficient is $r = 0$. There are two ways to evaluate your calculated r against this null hypothesis. One is to compute a t ratio, and the other is to use a table to look up the value.

The appropriate t ratio to use may be found in many statistics texts dealing with the product-moment correlation coefficient. It will not be discussed in this book.

Referring to the table involves knowing the sample size, N, and the significance level, α. Does the null hypothesis $r = 0$ call for a one-tailed or a two-tailed test of significance?

A. One-tailed
B. Two-tailed

(A, *go to* 46)
(B, *go to* 47)

---

46 The null hypothesis H_0: r = 0 can be rejected by a large enough positive r or a large enough negative r. That is, in the probability distribution of r, we will use r values in either tail as evidence for rejection of the null hypothesis.

(*Go to* 47)

47 The null hypothesis r = 0 requires a two-tailed test of significance. Table 5 in Appendix A gives the value of r required for significance. The degrees of freedom (df) for the distribution of r is expressed by N − 2. (One degree is lost by taking the variation about the mean $[(X - \overline{X})$ and $(Y - \overline{Y})]$, and the other results from only N − 1 cases or observations being needed to explain all variances, the last case being redundant.)

Look at Table 5 in Appendix A. With a sample of 52 cases (df = 52 − 2 = 50) it would take a computed r of .273 or greater to be significant at the .05 level when using a two-tailed test. That is, if the population value for the correlation coefficient was actually zero (the null hypothesis), only 5 times in 100 would you find $r \geqslant .273$ in a random sample from that population.

Referring back to the correlation computed in Table 11-1, where N = 10 and r = .971, determine the significance level for r for a two-tailed test.

A. $p < .0005$
B. $p < .001$
C. Neither of these

(A, *go to* 48)
(B, *go to* 49)
(C, *go to* 48)

48 You are dealing with a two-tailed test, and therefore the correct answer is $p < .001$. Follow carefully: N = 10, r = .971, df = 10 − 2 = 8. Entering Table 5 at the row df = 8, we find an r of .549 necessary for significance at the .05 level for one tail and at the .10 level for two tails. Our r of .971 is much greater than .549. Moving to the next column, we find an r of .632 necessary for significance at the .05 level, two tails. Again, our computed r is greater than the table r. We continue this procedure until the table r surpasses our calculated r. At no time does this happen for r = .971 (df = 8). Therefore our significance level is less than p = .001. We report the significance of our computed r of .971 at $p < .001$.

(*Go to* 49)

---

49 The correct answer for a two-tailed test is $p < .001$.

The null hypothesis H_0: $r = 0$ is *rejected* because our observed r is too large to be a chance result if r for the population really were zero. Therefore the population correlation must be not-zero. Note that the above was a two-tailed test. If our computed r was $-.971$, we still would have rejected the null hypothesis. Note also that we are not 99.9% confident that our population correlation is .971. Rather, we are 99.9% confident that our population correlation is not zero. (*Go to* 50)

---

50 The usual null hypothesis for a two-tailed test of r is H_0: $r = 0$.

A typical null hypothesis for a one-tailed test of r is H_0: $r \leqslant 0$.

Which hypothesis of interest (i.e., research hypothesis) below would call for the null hypothesis $r \leqslant 0$?

A. Blood pressure is positively related to obesity in United States men over 21 years old.

B. There is a relationship between alcohol consumption and age at death in women.

C. In Wistar strain rats, convulsion duration in seconds is inversely related to drug dosage of flurothyl in mg/100 ml.

(A, *go to* 51)
(B, *go to* 52)
(C, *go to* 53)

---

51 The hypothesis of interest stating that blood pressure and obesity are "positively related" calls for a test of a null hypothesis of "no relationship or no negative relationship." This is expressed as H_0: $r \leqslant 0$.

(*Go to* 54)

---

52 The statement "a relationship" implies that we are willing to accept either a positive or negative association. The null hypothesis, therefore, would be "no relationship" or $r = 0$.

Return and try again. (*Go to* 50)

---

53 The statement "inversely related" is our statement of interest or our research hypothesis. Inversely related is the same as negatively related. The alternatives that we wish to reject are, therefore, "no relationship" and "positive relationship." The null hypothesis to be tested for the inverse relationship between convulsion duration and mg/100 ml flurothyl is H_0: $r \geqslant 0$.

Return and try again. (*Go to* 50)

---

54 Recall that in discussing r, the product-moment correlation coefficient, we called it a descriptive statistic, but that it is commonly accompanied by an inferential act of (1) testing a hypothesis or (2) inferring a population value.

The best estimate of the population correlation is your sample correlation. Often, however, it is better to calculate a confidence interval for r rather than to make just a point estimate of r itself. The interval estimate takes into account the probability of error in your estimate and produces a range or interval for r at some predetermined level of confidence.

With a 95% confidence interval for r we can say that in the long run, we are "95% confident that the population correlation lies within our interval." Which statement below is a confidence interval?

A. (Estimated value) ± (2 standard deviations)
B. (Estimated value) ± (Z_α) (standard error of estimated value)

(A, *go to* 55)
(B, *go to* 56)

55 You seem to be confusing earlier discussions about the normal curve with concepts regarding confidence intervals. The standard deviation (related to deviations about the mean) is not the appropriate error term to relate to an interval about a statistic. The standard deviation will tell you something about the proportion of cases within an interval.

The 95% confidence interval is expressed as:

Estimated value ± ($Z_{.05}$) (Standard error of estimated value)

(*Go to* 56)

56 The confidence interval is essentially your estimate ± error term. The error term is defined as:

(Z_α) (Standard error of estimate)

Here we take Z at whatever α level we select. For example, for $\alpha = .05$, which represents the confidence level .95, the Z value from your Z table (areas under the normal curve, Appendix A, Table 2) is $Z_{.05} = 1.96$. (*Remember*, you are dealing with two tails in confidence intervals!)

Unfortunately, the above form for a confidence interval assumes a normal distribution for the estimated parameter. The distribution of sample r's tends toward nonnormality in many cases. So first r must be normalized, which is done by converting it to what is called a Z value. Then we determine a confidence interval for the resulting Z value and convert the resulting interval values back into r.

Turn to Table 6 in Appendix A; this represents the r to Z transformations. For example, if we had a sample size of 28 cases, computed an r of .68, and wanted to calculate a 95% confidence interval, we would perform the following steps.

1. In Table 6, find that r = .68 corresponds to Z = .829.
2. Standard error of Z = $\sigma_z = \frac{1}{\sqrt{N-3}}$, which in this example is

   $$\sigma_z = \frac{1}{\sqrt{28-3}} = \frac{1}{\sqrt{25}} = .2.$$
3. Compute confidence interval for Z as Z ± (Z_α) (σ_z) = .829 ± (1.96) (.2) = .829 ± .392 = .437 to 1.221.
   Note that $Z_{.05}$ for confidence intervals represents the area beyond .05/2 = .025 of the normal Z table (Appendix A, Table 2).
4. Convert Z_{lower} = .437 and Z_{upper} = 1.221 back into r by going down the Z column in Table 6, Appendix A, and finding the associated r value. Here Z of .437 is r of approximately .41 and Z = 1.221 is r = .84.
5. Our 95% confidence interval for r in the population is from .41 to .84. That is, in repeated samplings from the population, 95 times out of 100, our interval, here $.41 \leqslant r \leqslant .84$, would include the population correlation.

Now using the formula

$$\text{Confidence interval r} = Z \pm (Z_\alpha)\left(\frac{1}{\sqrt{N-3}}\right)$$

and the appropriate tables in Appendix A, calculate the 99% confidence interval for the correlation coefficient you computed from the data in Table 11-1.

$$N = 10,\ r = .971$$

When you are ready to check your work . . . (*Go to* 57)

---

57 $N = 10 \quad r = .971 \quad \alpha = .01$

1. In Table 6, we find that r = .971 corresponds to Z = 2.092.
2. α = .01/2 (for both sides of r) = .005, which corresponds to 2.57 found in Table 2.
3. Plugging these into the formula:

$$2.092 \pm 2.57\left(\frac{1}{\sqrt{10-3}}\right) = 2.092 \pm 2.57\left(\frac{1}{\sqrt{7}}\right) =$$

$$2.092 \pm 2.57\left(\frac{1}{2.65}\right) = 2.092 \pm 2.57\ (.38) =$$

$$2.092 \pm .98 = 1.112 \text{ to } 3.072$$

Continued.

Converting Z back to r in Table 6, we find $r_{lower} = .80$ and $r_{upper} = .99$. We are 99% confident that the correlation between self-image and reading achievement in our *population* of pupils is somewhere between .80 and .99. Or: $.80 \leq r_{pop} \leq .99$.

You have completed the subsection on the product-moment correlation coefficient. If you wish to start the frames on regression analysis . . .

(*Go to* 58)

If you wish to skip regression analysis and go to the rank-order correlation coefficients . . .

(*Go to* 81)

---

58 Within the concept of association we have discussed both correlation and prediction. *Regression analysis* is concerned with prediction. It is a method whereby data on scores from one sample may be used to predict scores on cases in a second sample.

We will be concerned only with the two-variable case in this book. Regression analysis involving more than two variables (called multiple regression analysis or multivariate regression analysis) is a slightly more advanced topic that can be found in many texts.

A good deal of the practical orientation of many professions and scholarly disciplines uses applied prediction. Diagnosis is essentially a prediction given certain measures of the given patient and based on past experience with those measures. Education uses tested techniques to achieve certain outcomes. Regression analysis tries to find some equation that best enables us to predict a Y score given an X score. Linear regression analysis tries to find the straight line through a set of data that best fits the data. In a scatter diagram, this is called the "best fit" line.

Recall from your high school or college algebra that the formula for a straight line is:

$$Y = a + bX$$

You will have to know this formula.

If we arbitrarily replace the constants a and b with 3 and 2, respectively, we would get the equation $Y = 3 + 2X$; for every X we can now calculate a Y. As X gets bigger, Y gets bigger. The value 2, which replaced the symbol b, is the *slope* of the line.

Look at the equation $Y = a + bX$ again. If Y gets bigger when X gets bigger, then the straight line has a (positive/negative) slope.

A. Negative
B. Positive

(A, *go to* 59)
(B, *go to* 60)

---

59 If Y gets bigger when X gets bigger, then in $Y = a + bX$, the value b must be *positive*. For Y to get smaller or more negative as X gets bigger, b would have to be a negative number. When b is negative, we say the slope is negative.

(*Go to* 60)

---

60 In $Y = a + bX$, if Y gets bigger when X increases in value, then b must be a positive number, so the slope is positive.

Consider two straight-line equations:

1. $Y = 2 + 3X$
2. $Y = 2 + 6X$

Also consider the plotting of those lines in Fig. 11-9. (A in Fig. 11-9 corresponds to equation 1, etc.) Note that in equation 1, for every unit increment in X, Y is incremented by $3(1) = 3$. In equation 2, for every unit increment in X, Y is incremented by $6(1) = 6$. We say that equation 2 has a *steeper* slope than equation 1.

In the equation $Y = a + bX$, which coefficient, a or b, indicates the slope?

A. a
B. b

(A, *go to* 61)
(B, *go to* 62)

FIG. 11-9

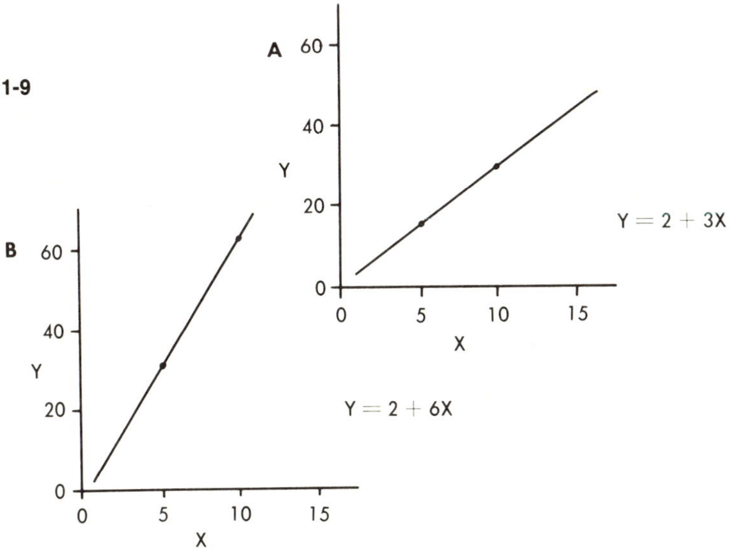

61 Remember that we spoke of b in the equation $Y = a + bX$ as indicating both the direction and the steepness of the slope. Therefore it is b that indicates the slope of the best-fit line.

Comparing the two equations

$$Y = 4 - 3X \qquad\qquad Y = 4 - 6X$$

we can say that both lines have negative slopes and that the equation on the right has a steeper slope than the equation on the left.

If this is not clear . . .

(*Go to* 58)

If you are reasonably clear on the concept "slope" . . .

(*Go to* 62)

---

62 b is the slope in the equation $Y = a + bX$. The slope indicates both direction and steepness.

In the general equation for the straight line, $Y = a + bX$, what is the value of Y when $X = 0$?

A. Unknown
B. a

(A, *go to* 63)
(B, *go to* 64)

---

63 Although it may appear confusing because you are mixing letters (Y, a, b, and X) and numbers, the value of Y when $X = 0$ is:

$$Y = a + bX = a + b(0) = a$$

(*Go to* 64)

---

64 In $Y = a + bX$, if X is zero, $Y = a$. The value of a is called the Y intercept because it is the point on the Y axis at which the line intercepts the Y axis. In Fig. 11-10, depicting the regression equation $Y = -3 + 2X$, note that the line intersects Y below the X axis at $Y = -3$.

Which equation below describes a straight line with a positive slope, a steepness of 7, and a Y intercept of 2?

A. $Y = 2 - 7X$
B. $Y = 2 + 7X$
C. $Y = 7 + 2X$

(A, *go to* 65)
(B, *go to* 67)
(C, *go to* 66)

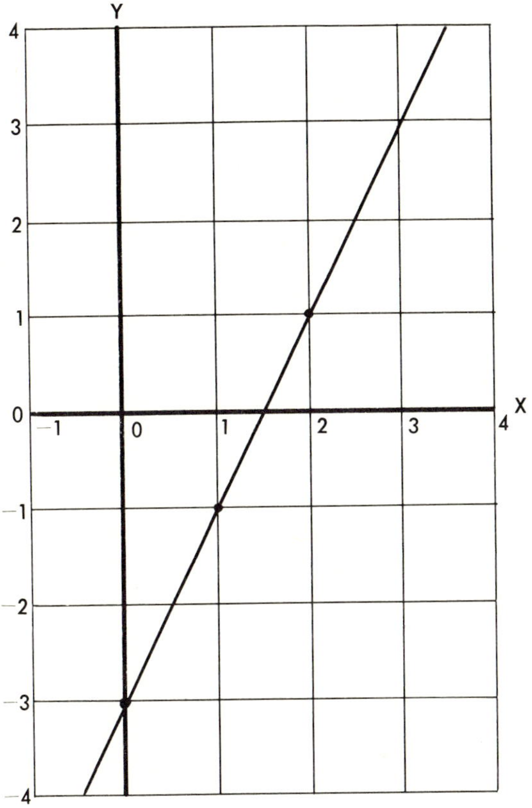

FIG. 11-10. Regression line for $Y = -3 + 2X$.

---

65 The description calls for a regression equation for a straight line with a positive slope. The equation you chose ($Y = 2 - 7X$) has a negative slope. The number -7 replaced the symbol for slope (b); to obtain a positive slope, the sign in front of the slope must be positive.

(Go to 67)

---

66 The equation you chose ($Y = 7 + 2X$) reverses the two concepts of slope and intercept. The generic equation for the straight line is $Y = a + bX$ where a = the Y intercept and b = the slope. Try that frame again.

(Go to 64)

---

67 In $Y = 2 + 7X$, 2 is the Y intercept and 7 is the slope.

Turn to Fig. 11-11 and plot the straight line $Y = 2 + 7X$. So that you can more accurately check your results, use $X = 3$.

When you are ready to check your work, look at Fig. 11-12.

When you are ready to go on . . . (*Go to* 68)

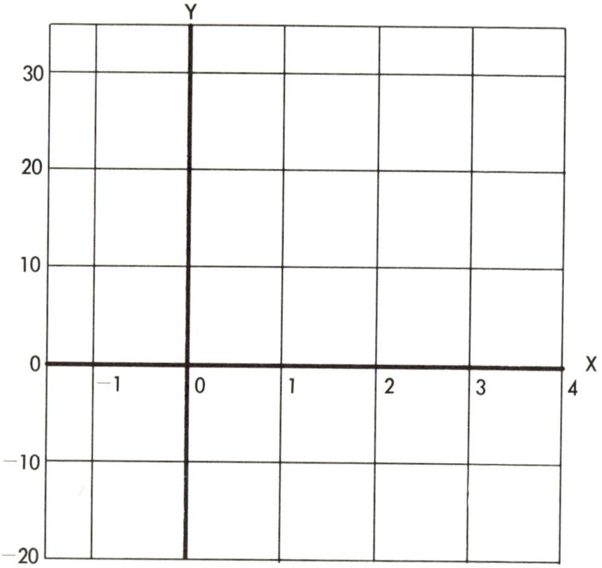

Fig. 11-11. Plot for $Y = 2 + 7X$ (I).

68 Solving the regression equation $Y = 2 + 7X$ when $X = 0$:

$$Y = 2 + 7(0)$$
$$Y = 2 + 0$$
$$Y = 2$$

Then, solving when $X = 3$:

$$Y = 2 + 7(3)$$
$$Y = 2 + 21$$
$$Y = 23$$

That is, when $X = 0$, $Y = 2$, and when $X = 3$, $Y = 23$. Given the two points, you should be able to draw a straight line in Fig. 11-12 to represent the equation $Y = 2 + 7X$.

If X is the grade of a student in elementary school and Y is the score on a standardized test for concept difficulty, what score would you predict as the norm for a student in the fourth grade?

When you have found the answer . . . (*Go to* 69)

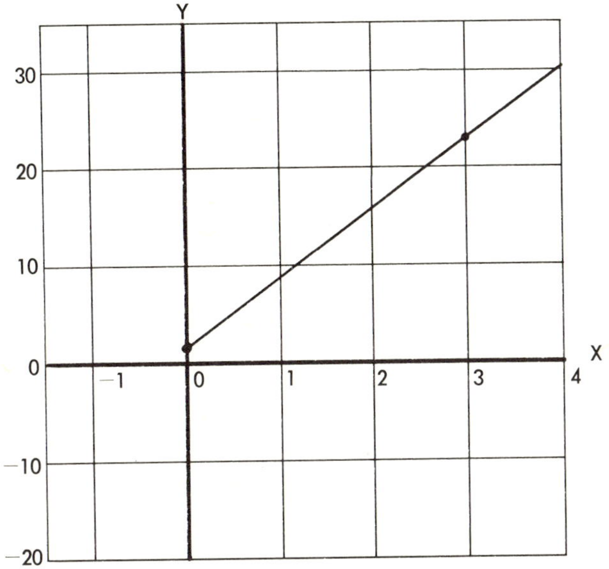

FIG. 11-12. Plot for Y = 2 + 7X (II).

69 By the time a student is in the fourth grade, that student should have achieved a concept difficulty score of 30. You could have arrived at that answer in two ways:

1. You could have solved the equation

$$Y = 2 + 7X = 2 + 7(4) = 2 + 28 = 30$$

2. You could have simply looked at the plot that you drew (Fig. 11-11) and determined the value of Y when X = 4

In the form Y = a + bX we say that we are predicting Y, given X. We also say that the value of Y is dependent on the value of X, or Y is the *dependent* variable. In the same way we consider the variable X to be free to vary, thus producing different Y values. X is said to be independent of Y, or X is the *independent* variable. (Note: X does not *cause* Y. We are simply labeling the two variables.)

Which variable is the dependent variable in the following problem? "The researchers wish to use age of respondent at death of spouse to predict life expectancy in years after death of spouse."

A. Life expectancy
B. Respondent's age
C. Death of spouse

(A, *go to* 70)
(B, *go to* 71)
(C, *go to* 72)

70 Life expectancy is the correct dependent variable. The terms "predictor variable" and "independent variable" mean the same. Similarly, the "dependent variable" is sometimes called the "criterion variable" because it is the criterion of success or interest that we wish to predict.

"A regression analysis of data from a sample of 162 men produced an equation for using social class of parents to predict probability of police detention during the ages 13 to 18."

The independent variable is ________________. (*Go to* 73)

---

71 Respondent's age is not the dependent variable. Remember that the variable we are trying to predict *depends* on the value of the variable we are using for prediction. Y depends on X in Y = a + bX.

Now reanalyze the problem: "The researchers wish to use age of respondent at death of spouse to predict life expectancy in years after death of spouse." The variable to be predicted is "life expectancy in years after death of spouse." Y = life expectancy. The variable we are using for prediction is "age of respondent at death of spouse." X = age of respondent.

Y is the dependent variable.

X is the independent variable. (*Go to* 70)

---

72 Death of spouse is not a variable in the regression equation. Any respondent who was never married or whose spouse is still alive is not a member of the population or sample that we are dealing with.

(*Go to* 71)

---

73 The independent variable is social class of parents. The equation to be produced is:

Probability of police detention = a + b(Parental social class)

This is a linear equation. Remember that if the data are curvilinear, a linear equation will not produce an accurate prediction. There are three options for handling curvilinear data.

1. Sometimes converting scores into logarithms will straighten out a curve.
2. Break the X variable into two or three intervals and calculate separate equations for each part.
3. Consult more advanced texts on statistics for these techniques.

Before considering how to calculate the two constants, a and b, in the regression equation, a quick review is appropriate.

1. The general linear form is Y = a + bX.
2. a is the Y intercept (where the line crosses the Y axis).
3. b is the slope (direction and steepness) of the line.
4. Y is the dependent, criterion, or predicted variable.
5. X is the independent or predictor variable.
6. Before you can get good linear regression prediction, the data must lie near a straight line. Random data or curvilinear data cannot produce good linear prediction.

(*Go to* 74)

---

74 Next consider how the linear equation is determined. We have observed X scores and Y scores. We want an equation to *predict* Y scores given the observed X scores. If we let Y stand for the observed score, then we *predict a* Y score (call it $\hat{Y}$, pronounced "Y hat"). $\hat{Y} = a + bX$. The difference between Y (observed) and $\hat{Y}$ (predicted) for each person is our error in prediction.

For any X, our predicted value, $\hat{Y}$, is the point on the regression line presented in Fig. 11-13. We want to minimize our error, that is, the difference between Y and $\hat{Y}$, or we want $(Y - \hat{Y})$ = minimum.

(*Go to* 75)

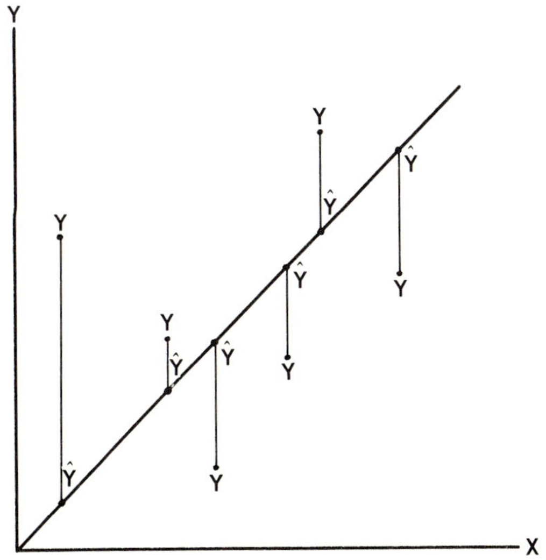

FIG. 11-13. $\hat{Y} = a + bX$.

75 Actually, what we minimize is not $(Y - \hat{Y})$ but $(Y - \hat{Y})^2$. Just as in the standard deviation, we work with the squared deviation. Squaring the error does several things: it makes a unique mathematical solution possible; it eliminates the negative value problem; and it gives added weight to the effect of those points farther away from the predicted value.

To minimize $(Y - \hat{Y})^2$ is to use what is known as the *least squares* solution.

As a matter of interest, you might note that our set of predicted values ($\hat{Y}$, i.e., those on the regression line) will generally lie closer to the mean of the Y variable ($\bar{Y}$) than our set of observed values (Y). That is why the best fitting prediction line is called a "regression equation"; the predicted values tend to regress toward the mean.

Finally, note that the regression equation for Y predicted by X is not the same as X predicted by Y. For instance, in the two equations

$$Y = a_{yx} + b_{yx}X \qquad\qquad X = a_{xy} + b_{xy}Y$$

the value b_{yx} is not equal to b_{xy}.

If you would like to go over these last few frames again . . .

(*Go to top of p.* 263)

If you are ready to proceed to calculations . . .

(*Go to* 76)

76 Our concern is with obtaining the two constants in the regression equation, a and b. Since we are predicting Y ($\hat{Y}$) from our knowledge of X and Y (the empirically observed values), the regression equation becomes

$$\hat{Y} = a + bX$$

$$b = \frac{\Sigma(X - \bar{X})(Y - \bar{Y})}{\Sigma(X - \bar{X})^2} = \frac{N\Sigma XY - (\Sigma X)(\Sigma Y)}{N\Sigma X^2 - (\Sigma X)^2}$$

$$a = \bar{Y} - b\bar{X} = \frac{\Sigma Y}{N} - b\frac{\Sigma X}{N}$$

If you think these equations seem similar to those used in computing the product-moment correlation coefficient (r), you are right. The slope b (sometimes called the regression weight or regression coefficient) is closely related to r.

$$b = r\frac{\sigma_y}{\sigma_x} \text{ and } r^2 = b_{yx}b_{xy}$$

where σ = the standard deviation of each of the variables.

Return to Table 11-1 (the data relate self-image and reading achievement scores for first-graders), and, using the formulas just given and the calculations that you did earlier, calculate b and a.

$$b = \frac{N\Sigma XY - (\Sigma X)(\Sigma Y)}{N\Sigma X^2 - (\Sigma X)^2}$$

$$a = \frac{\Sigma Y}{N} - b\frac{\Sigma X}{N}$$

When you are ready to check your calculations . . .

(*Go to* 77)

77

$$b = \frac{N\Sigma XY - (\Sigma X)(\Sigma Y)}{N\Sigma X^2 - (\Sigma X)^2} = \frac{10 \cdot 440 - (61)(60)}{10 \cdot 455 - (61)^2} = \frac{4400 - 3660}{4550 - 3721} = \frac{740}{829} = 0.893$$

$$a = \frac{\Sigma Y}{N} - b\frac{\Sigma X}{N} = \frac{60}{10} - 0.893\frac{61}{10} = 6 - 0.893(6.1) = 6 - 5.45 = 0.55$$

What is the linear regression equation for reading achievement predicted from self-image?

A. Reading achievement = 0.893 + 0.55(self-image)
B. Reading achievement = 0.55 + 0.893(self-image)
C. Self-image = 0.55 + 0.893(reading achievement)

(A, *go to* 78)
(B, *go to* 79)
(C, *go to* 80)

---

78 You chose reading achievement = 0.893 + 0.55(self-image), which is the equation $\hat{Y} = b + aX$. Try again.

(*Go to* 77)

---

79 The correct form of the equation is reading achievement = 0.55 + 0.893(self-image). Note that when the child's self-image score = 2, the reading achievement score ($\hat{Y}$) =

$$0.55 + 0.893(2) = 0.55 + 1.786 = 2.336$$

In the data in Table 11-1, the two cases with self-image of 2 were associated with reading achievement scores of 3 and 2. Using our equation, we would predict a reading achievement score of 2.336, which is not far from the observed scores. Because we have used a regression equation to predict these scores, we are slightly off the observed values, but we are able to predict all Y values from this single equation.

You have completed the section on linear regression analysis for two variables.

(*Go to* 81)

---

80 You chose self-image = 0.55 + 0.893(reading achievement), which is the equation $\hat{X} = a + bY$. Try again.

(*Go to* 77)

---

81 *Rank order correlation* (sometimes called rank-difference correlation) can be considered as a special case of the product-moment correlation. It may be applied to any paired data (bivariate) that can be ordered from low score to high score. Thus it can be used with which scale of measurement?

A. Nominal scale or higher
B. Ordinal scale or higher
C. Interval scale or higher

(A, *go to* 82)
(B, *go to* 83)
(C, *go to* 84)

---

82 Nominal scales are really nothing more than named categories that cannot be ordered in any high-low sense. It takes at least an *ordinal* scale to have order in the measures so that the scores can be ranked.

(*Go to* 83)

---

83 In addition to being usable with ordinal data, rank order correlation is useful for interval data when you are unwilling to assume that the two variables are normally distributed in the population. The product-moment correlation, r, assumes this type of normality. Nonnormality occurs frequently, although r is by far the most commonly encountered measure of association.

(*Go to* 85)

---

84 Although you can use a rank order correlation with an interval scale or higher, you can also use it with an ordinal scale, since scores on an ordinal scale can be ordered or ranked.

(*Go to* 83)

---

85 If we had four scores (70, 82, 78, 65) and wanted to rank them, we would list them thus:

Score	*Rank*
82	1
78	2
70	3
65	4

On a separate sheet of paper, place in order and assign ranks to the following data:

36, 48, 24, 25, 42, 41, 30

When ready . . .

(Go to 86)

---

86 Your ranking should look like this:

Score	*Rank*
48	1
42	2
41	3
36	4
30	5
25	6
24	7

Now figure out what rank you would assign to persons b and c in the table below:

Person	*Score*	*Rank*
a	97	1
b	83	2.5
c	83	2.5
d	79	4
e	76	5

A. 2, 3
B. 2.5, 2.5

(A, *go to* 87)
(B, *go to* 88)

---

87 By assigning the ranks 2 and 3 to persons b and c you are arbitrarily giving one of the individuals a higher rank than the other.

(Go to 88)

---

88 When there are tied scores, you assign each of the tied scores the *mean of the rank positions* that they occupy. In this case both persons b and c get the mean of rank positions 2 and 3, or $\frac{2+3}{2} = 2.5$.

Person	*Score*	*Rank*
a	97	1
b	83	2.5
c	83	2.5
d	79	4
e	76	5

Continued.

Now that we can assign ranks to scores, we can compute a correlation between two ranks. The rank order correlation to be discussed here is called the Spearman rank order correlation coefficient (developed by a man named Spearman). The statistic is symbolized usually as ρ (the Greek letter rho) or as $r_{(rho)}$. Rho is concerned with the difference in ranks for two separate variables being correlated for each individual or observation.

Table 11-2 presents data from a sample of seven respondents who were top administrators in a government agency. The administrators were given ratings by their supervisor on leadership ability and by their co-workers on friendliness. Since the ratings represent only ordinal-level variables, we will use the rank order correlation coefficient to determine whether their positions on the two variables are related.

On a separate sheet of paper or in Table 11-2, write down the ranks for each person on the two variables (rank from high to low).

When you have done this . . .

(*Go to* 89)

TABLE 11-2. Relationship of leadership ability and friendliness

Person	*Leadership ability*	*Leadership rank*	*Friendliness*	*Friendliness rank*	*D*	D^2
A	3		11			
B	5		12			
C	6		15			
D	5		8			
E	2		6			
F	4		4			
G	1		2			

89 The ranking for the two variables should be as follows:

Person	*Leadership rank*	*Friendliness rank*
A	5	3
B	2.5	2
C	1	1
D	2.5	4
E	6	5
F	4	6
G	7	7

If you ranked from low to high, your ranks would appear different, but it will not matter in the correlation.

The column headed D in Table 11-2 is used for the difference between the two ranks of an individual. For person A, the difference in ranks D = 5 − 3 = 2. What is D for person D?

A. −3
B. −1.5

(A, *go to* 90)
(B, *go to* 91)

---

90 You answered that D = −3, which you may have calculated by finding the difference between the raw score for leadership ability (5) and friendliness (8). The difference between ranks (D) is 2.5 − 4 or

________________.

(*Go to* 91)

---

91 The difference between ranks (D) for person D = −1.5.

Now finish calculating D and D^2 for Table 11-2. When you are done . . .

(*Go to* 92)

---

92 Your table should now read as follows:

Person	*Leadership ability*	*Leadership rank*	*Friendliness*	*Friendliness rank*	*D*	*D²*
A	3	5	11	3	2	4
B	5	2.5	12	2	0.5	0.25
C	6	1	15	1	0	0
D	5	2.5	8	4	−1.5	2.25
E	2	6	6	5	1	1
F	4	4	4	6	−2	4
G	1	7	2	7	0	0

You are now ready to calculate the rank order correlation. The formula for the Spearman rank order correlation is

$$\rho = 1 - \frac{6\Sigma D^2}{N(N^2 - 1)}$$

where

D = difference between pairs of ranks
N = number of pairs of ranks

Using the data in Table 11-2, calculate ρ. When you are finished . . .

(*Go to* 93)

93

$$\rho = 1 - \frac{6\,(11.5)}{7(7^2 - 1)} = 1 - \frac{69}{7(49 - 1)} = 1 - \frac{69}{7(48)}$$

$$= 1 - \frac{69}{336} = 1 - .20 = .80$$

Or, with $\rho = .80$, it is fairly safe to say that leadership ability and friendliness are fairly highly related, and in a positive direction; i.e., as leadership ability as rated by supervisor increases, the rating of co-workers on a friendliness dimension also increases.

What would be the value of rho if every person in the sample had the same rank on variable X as on variable Y?

A. 0.0
B. 1.0

(A, *go to* 94)
(B, *go to* 95)

---

94 Consider what D would be for each person if that person had the same rank on both variables. Each D would be zero (e.g., 5 − 5 = 0); that is, there is no difference in rank. The term $6\Sigma D^2$ would thus be zero, and dividing anything into zero gives zero. Thus rho would be 1.0 − 0 = 1.0.

(*Go to* 95)

---

95 If all persons have the same rank on both variables, $\rho = 1.0$.

If the two rank orders display a perfect inverse relationship to each other, that is, if as the rank order on one variable increases, the rank order on the other variable decreases, then rho will = −1.0. Thus the limits, and indeed the interpretation, of rho are the same as for the product-moment correlation.

One note of caution is needed. If a large proportion of the scores are ties, it is appropriate to use a correction formula for rho. Consult an advanced text for help in these cases.

The basic inferential statistical test for the descriptive statistic rho is the same as for r, namely:

$$H_0\!: \rho = 0$$
$$H_A\!: \rho \neq 0$$

Note that this is a two-tailed test (rho can be not-zero by being positive or negative). The probability associated with the test indicates the likelihood of finding your observed sample rho if the population correlation was actually zero. Table 7 in Appendix A gives the value of rho required for significance for various small sample sizes (N = number of pairs of ranks).

With ten pairs of ranks (ten cases) it would take a rho of ±.648 or greater to be significant at the .05 level for a two-tailed test. That is, if the population rho was actually zero, only 5 times in 100 would one find a rho of ±.645 or larger in a random sample from that population.

Using the data from Table 11-2, the value of rho that you calculated for it ($\rho = .80$), and the two-tailed hypothesis above, determine the significance level.

A. $p > .05$ (A, *go to* 96)
B. $p < .05$ (B, *go to* 97)
C. $p < .01$ (C, *go to* 98)

---

96 The value of rho calculated was .80. There are seven pairs of ranks. Entering Table 7 of Appendix A at N = 7, we find a rho of .786 or greater necessary to be significant at the .05 level for two tails. Since .80 > .786, the probability of our observed rho is <.05. Our observed rho, however, is smaller than that necessary for significance at the .01 level (.929). Therefore $p < .05$ but > .01. (*Go to* 97)

---

97 For $\rho = .80$, N = 7, two tails, $p < .05$ but > .01.

If our null hypothesis was that there was no correlation between the two variables, should our null hypothesis be rejected or not rejected at the .05 level?

A. Not rejected
B. Rejected

(A, *go to* 99)
(B, *go to* 100)

---

98 The value of rho calculated was .80. There are seven pairs of ranks. Entering Table 7 of Appendix A at N = 7, we find a rho of .786 or greater necessary to be significant at the .05 level for two tails. Since .80 > .786, the probability of our observed rho is < .05. Our observed rho, however, is smaller than that necessary for significance at the .01 level (.929). Therefore $p < .05$ but > .01. (*Go to* 97)

---

99 Since $p < .05$ for $\rho = .80$, then a rare event has occurred in our sample that would only occur 5 times or less out of 100 *if in fact the true rho were zero*. The event (rho as large as .80) is so rare that we decide the true rho cannot actually be zero. That is, we reject the null hypothesis of no correlation (H_0: $\rho = 0$). (*End of chapter* 11)

---

100 The null hypothesis should be *rejected;* your answer is correct.

(*End of chapter* 11)

Examination for chapters 6 to 11

To test the adequacy of your understanding of the content material to this point in the book, it is recommended that you take the following examination. The examination was devised by turning most of the learning objectives at the beginning of each chapter into questions. If you met the objectives outlined, then you should have no trouble with the examination.

If the examination question requires the calculation of a statistic, look the formula up in the text or in Appendix C; otherwise, do not use your text. Appendix B contains the answers to the questions. If you answered a question incorrectly, the answer is referenced to a specific chapter and learning objective. You should go back to that chapter and the appropriate frames to restudy that particular material.

GOOD LUCK

1. What is the probability of an event that will:
   A. Absolutely happen?
   B. Absolutely not happen?
2. A. In a jar filled with plastic balls, there are ten white, ten black, and ten blue balls. What is the probability of getting a white or blue ball in one draw from the jar?
   B. In two draws from that jar, without replacement, what is the probability of getting a white ball and then a black ball?
3. Which of the following events can be classified as independent?
   A. The random selection of one man and one woman from a class of students.

B. Draw two random samples of ten each from a population of one hundred.
C. Six dice are tossed simultaneously producing each of the six sides, that is, die 1 displays a 1, die 2 displays a 2, etc.

4. Which of the following statements best describes internal validity?
   A. Ensuring that the experimenter follows the rules.
   B. Ensuring that an experiment is reasonably representative of reality.
   C. Ensuring that the results were really a function of the experimental treatment.
5. Which of the statements in question 4 best describes external validity?
6. Two groups of weight-watchers start meeting weekly after Christmas. One group leader is using transactional analysis to increase the sensitivity of the individual, the other group leader is using a token-reinforcement method for lost weight. At the end of spring the second group has lost more weight.

   Which of the following statements represent threats to internal validity and which represent threats to external validity in the experiment?
   A. No attempt was made to establish group equivalence.
   B. Weight loss was due to season of year.
   C. Members of the groups were not representative of all persons who try to lose weight.
   D. Leaders of the groups differ in style and personality.
7. Aesop observed that when the fox was unable to reach the grapes, he declared them to be sour anyway. More recently, similar results in experiments led a researcher (Festinger, 1957) to conclude that when a person is prevented from reaching a desired goal by his own inability, cognitive dissonance is reduced by reducing the value of the goal in his mind.

   Was the hypothesis arrived at by a process of deduction or induction?
8. Hypothesis: Holding age and education constant, nurses with disaster training will be rated more favorably by patients at the end of a real disaster episode than nurses without disaster training. Identify the variables:

   A. Independent variable: ______________________________

   B. Intervening variable: ______________________________

   C. Control variable: ______________________________

   D. Dependent variable: ______________________________

9. Ms. Brown observes that children in Head Start programs get considerable practice in reading readiness skills. Since she knows that practice improves these skills, she concludes that those children who have participated in Head Start will be superior in reading readiness.

   Was the hypothesis arrived at by a process of deduction or induction?

10. Of the following, the most important determinant of the extent to which the results of a research study can be generalized is the:
    A. Nature of the sample used in the study.
    B. Adequacy of the instruments used to collect data.
    C. Degree to which the effects of extraneous variables have been controlled.
    D. Ease with which the research design can be replicated.
11. The most efficient means of ruling out such factors such as maturational level, ability, and age as rival explanations of an observed posttest difference between an experimental and a control group is:
    A. Careful matching of the experimental and control groups on all possible factors such as sex, IQ, age, etc.
    B. Using a time-series quasi-experimental design.
    C. Random assignment of subjects to the treatment groups.
    D. Administering a pretest.
12. Match the two terms "validity" and "reliability" with the following statements:
    A. Do you get the same results on repeated measures; is it consistent?
    B. Does it measure what is claimed for it?
13. If you are taking a multiple choice examination with five questions where each question has three choices and if you guess on each of the five questions, what is the most probable number of questions that you will get correct?
14. Any statistic (e.g., mean, variance, median) has a theoretical distribution if we assume that we took a large number of samples of a given size and computed the statistic for each sample. This distribution of the statistic is called the (A) distribution. The standard deviation of this distribution of the statistic is called the (B).
15. Which of the following statements meets the definition of a hypothesis?
    A. There is no difference between the mean incomes of poverty families in Gary, Indiana, and in Kansas City, Missouri.
    B. IQ scores differ in the children of the above two groups.
    C. The proportion of the population which is black in Gary is not at all similar to that in Kansas City.
    D. The proportion of blacks who are in-migrants to Kansas City is the same as that in Gary.

16. A. For the following research hypothesis, what would the appropriate null hypothesis be?
    H_A: IQ scores of poverty level children in Gary are lower than IQ scores of poverty level children in Kansas City.
  B. For the following null hypothesis, state two possible alternative hypotheses.
    H_0: The proportion of blacks who are in-migrants to Kansas City is the same as that in Gary.

17. Define:
  A. Type I (α) error.
  B. Type II (β) error.
18. As the probability of a Type I error increases, the probability of a Type II error __________________.
19. If you reject a null hypothesis at $\alpha = .01$, what is the probability that you will be incorrect in rejecting that null hypothesis?
20. Which form below is the correct generic formula for basic inferential tests of statistical significance?

A. $\dfrac{\text{Standard error of statistic} - \text{Expected statistic}}{\text{Obtained statistic}}$

B. $\dfrac{\text{Obtained statistic} - \text{Expected statistic}}{\text{Standard error of statistic}}$

C. $\dfrac{\text{Obtained statistic} - \text{Standard error of statistic}}{\text{Expected statistic}}$

21. For each of the following problems, select the appropriate test of statistical significance from the following list.

Tests

1. Z, single sample, $\bar{X}$
2. Z, single sample, proportion
3. t, single sample, $\bar{X}$
4. Z, two independent samples, $\bar{X}$
5. Z, two independent samples, proportion
6. t, two independent samples, $\bar{X}$
7. χ^2

  A. A researcher obtains a random sample of the Public Health Service venereal disease clinics across the country (n = 25). The P.H.S. is interested in ascertaining if the clinics are meeting the standard of performance established several years ago by a special commission on venereal disease. The commission determined that a standard score of 60 (a scale using multiple factors added together to produce a mean of 60) should be maintained by the P.H.S. to be considered successful in stopping epidemics of venereal disease. The researcher's sample produces a mean of 52 and a standard deviation of 12. Does the researcher have evidence to indicate that the V.D. clinics are meeting/not meeting the standard established as the norm of performance by the special commission? Which is the appropriate test?

B. You are interested in attempting to find out the attitudes held by a community's power structure regarding the effect of a reduction in Federal funds available for poverty programs. You suspect that individuals who are highly dogmatic are more in favor of local communities funding their own poverty programs than are those who score low on a dogmatism scale. To test this hypothesis you randomly sample twenty persons designated as members of the community power structure. You give them a questionnaire on which you ascertain (1) attitude toward federal funding of poverty programs and (2) level of dogmatism. Since dogmatism is your independent variable, you split the group of twenty at the median dogmatism score into high and low dogmatism groups. The data on attitudes toward federal funding of poverty programs is based on an interval scale ranging from 6 (= highly favorable) to 1 (= highly unfavorable). Given the following data, you want to test whether there is a statistically significant difference between the two dogmatism groups on the federal funding question:

Low dogmatism	*High dogmatism*
$N = 10$	$N = 10$
$\overline{X} = 4$	$\overline{X} = 2.8$
$s = 1$	$s = 1.47$

Which is the appropriate test?

C. In random samples of 25 female and 25 male high school students, a researcher asks each student to designate his/her anticipated social class in 10 years as being high, medium, or low. The researcher finds that of the boys, 6 say high, 12 medium, and 7 low. Of the girls, 15 say high, 7 medium, and 3 low. Is there a difference between male and female high school students and their anticipated social class? Which is the appropriate test?

D. A salesman for a drug company claims that a certain patent medicine marketed by his company is 90% effective in relieving pain from arthritis. His claim is based on a random sample of 200 people who have a confirmed diagnosis of arthritis. In this sample, the medicine provided relief for 160 people. Was the salesman's claim a legitimate one? Which is the appropriate test?

E. A drug company's detail man has presented a physician with data on controlling malignant hypertension (severe high blood pressure) when using its new F.D.A. approved drug, Xpheno, as opposed to using the standard drug of choice, Ypheno. The mean diastolic pressure in a random sample of patients with malignant hypertension after the use of Xpheno for a 2-month period is 86 ($s = 7$, $N = 100$), whereas Ypheno produces a mean diastolic pressure of 88 ($s = 6$, $N = 100$) for the same treatment duration. All controls appear to be adequate in their description of the clinical trials. Is there adequate statistical evidence to convince the physician to switch to the new drug? Which is the appropriate test?

F. In an attempt to determine the most effective method of teaching statistics to undergraduates, a teacher divides a class into two equal parts with 100 students in each of two groups. In group I the students are taught using traditional classroom methods, whereas in group II no lectures are given; instead students are exposed to a self-learning text covering the same content as the other group. The entire course is graded on a pass-fail basis using the same standard examination for both groups. The teacher determines that in group I, 80 students pass, whereas in group II, 90 students pass. If the teaching method is a factor in learning, which of the two methods should be recommended for use next year? Which is the appropriate test?

G. A social scientist working for the State Department of Mental Health collected data on the IQ's of mentally retarded children and their father's incomes. The state mean income for that year was estimated to be $8,206 with a standard deviation of $1,600. The sample of 120 children have fathers with a mean income of $7,811. Is there evidence here to indicate that the income of father's with retarded children differs from the overall state income? Which is the appropriate test?

22. A. For the problem stated in 21A, calculate the appropriate test statistic and reach a conclusion about the question asked at $\alpha = .05$.

B. For the problem stated in 21B, calculate the appropriate test statistic and reach a conclusion about the question asked at $\alpha = .05$.

C. For the problem stated in 21C, calculate the appropriate test statistic and reach a conclusion about the question asked at $\alpha = .05$.

D. For the problem stated in 21D, calculate the appropriate test statistic and reach a conclusion about the question asked at $\alpha = .05$.

E. For the problem stated in 21E, calculate the appropriate test statistic and reach a conclusion about the question asked at $\alpha = .05$.

F. For the problem stated in 21F, calculate the appropriate test statistic and reach a conclusion about the question asked at $\alpha = .05$.

G. For the problem stated in 21G, calculate the appropriate test statistic and reach a conclusion about the question asked at $\alpha = .05$.

23. For the sample mean in problem 21G, construct a 95% confidence interval and interpret the outcome.

24. Which of the following questions would call for a correlation procedure and which a prediction procedure?

    A. For all premature births, is there a relationship between mother's family's income and the number of weeks the birth of her first infant was premature?

    B. For all premature births, if a mother belongs to a family with an income of $6,000, how many weeks premature is the birth of her first infant likely to be?

25. In which of the following problems can a measure of association be calculated, and what would be the appropriate measure (r or ρ)?

    A. The self-esteem rating (high, medium, low) of men in the 1955 birth cohort with their perceived social class (high, medium, low).

    B. The College Entrance Examination scores of all men and women in the birth cohort of 1955.

    C. The College Entrance Examination scores of all women in the 1955 birth cohort with their grade-point averages.

    D. The College Entrance Examination scores of all men in the 1955 birth cohort.

26. Turn to Fig. 11-6 and determine whether a linear model is appropriate for the data plotted in the lower right-hand corner of that figure.

27. What would you estimate the correlation coefficient to be of the scatter diagram in Fig. 11-6? Your answer should be within $\pm .3$.

28. In a random sample of premature births (based either on birth weight or gestation) there was an interest in the relationship between the yearly income of the family of the new mother and the number of weeks premature. It was hypothesized that there would be an inverse relationship between income and prematurity. The following data were collected from records in a community hospital.

Subject	X Income in \$1,000	Y Weeks premature
1	8	0
2	6	1
3	4	5
4	4	10
5	5	3
6	4	7
7	7	1
8	5	2
9	6	4
10	9	0

A. Compute the correlation coefficient between income and weeks premature.
B. Test a hypothesis for the significance of the r that you just calculated at $\alpha = .05$. Interpret your results.
C. Calculate the 95% confidence interval for the r calculated above.
D. Compute the equation for a straight line using the data above.
E. If a mother's family's income is \$3,000 and she delivers prematurely, how many weeks premature would you predict her to be?

29. In a random sample of eight students, a psychologist is interested in ascertaining whether there is a relationship between the frequency of headaches reported during a randomly selected month and the student's ranking in the psychology class. To test this idea, he collects the following data:

Individual	Frequency of headaches for month	Class rank
1	3	40
2	5	61
3	0	7
4	6	70
5	1	21
6	3	12
7	2	19
8	4	43

A. Calculate the rank order correlation for the relationship between headache frequency and class rank.
B. Determine the significance of rho and interpret the results.

chapter 12

Evaluating the research report

At the conclusion of this chapter, you should be able to:

1. Identify in a scientific article the primary procedures used to guard against internal and external invalidity.
2. Identify and describe the sources of invalidity that a control group is designed to overcome.
3. Given a research article, determine (a) the objectives, hypotheses, or questions to be answered and (b) the control or comparison group used, if any.
4. Determine and specify whether conclusions in a research article can be justified by the findings and whether the conclusions are relevant to the objectives, hypotheses, or questions posed in the article.

---

1 This chapter deals briefly with sources of error in research and questions that may be put to a research report to evaluate its conclusions. In chapter 7 we suggested two kinds of validity in research. They were:

A. Internal and external validity
B. Statistical and sampling validity

(A, *go to* 2)
(B, *go to* 3)

---

2 Internal and external validity were discussed in chapter 7.

(*Go to* 4)

---

3 Statistical and sampling validity were not categories used in chapter 7. Instead the two categories of internal and external validity were discussed.

(*Go to* 4)

---

4 A study has *internal validity* when its outcomes can be related to the variables being tested rather than to other extraneous causes; that is, a study has internal validity when the treatments are known to have affected or not affected the results.

A study has *external validity* when one is confident the results would apply to specific other programs, populations, or situations.

Match both types of validity with the appropriate definitions that follow:

Internal validity—

External validity—

A. Can the results be generalized?

B. Is there a result?

(*Go to* 5)

---

5 The correct match would be:

Internal validity—Is there a result?

External validity—Can the results be generalized?

The questions to be posed are:

A. Does this study show a result, or are the findings explainable as being probably due to some extraneous factor or some error?

B. Where else can the findings apply, or is the sample unique and unlike others in which we are interested?

Other questions must be put to a research report in addition to asking whether there are plausible rival hypotheses. Questions regarding clarity and statistical soundness must be raised, for instance. In the following frames we will present briefly some factors to consider in judging the validity of a piece of research.

(*Go to* 6)

---

6 *Sources of bias from sampling error* primarily affect external validity. Sampling from a restricted population (e.g., those available as volunteers or convenience samples) or populations that change over time or space limit our ability to generalize. The group studied, in these cases, becomes unlike the group to which we wish to generalize. Examples might include different populations encountered in a day clinic versus a night clinic, those at home for an interview versus those not at home, volunteers

Continued.

versus randomly selected persons, or people in New York versus people in Denver.

Another source of bias in the sample comes from loss of cases in the sample, sometimes called *sample mortality*. If a control and experimental group suffer different kinds of losses over time, the two initially similar groups may become dissimilar. This difference is a potential source of internal invalidity, since different outcomes may be due to differences in the groups rather than treatment effects. If a randomly selected treatment group suffers losses in some systematically biased way, then external validity suffers. We cannot be sure the same results would occur in the population. Clinical studies often suffer nonrandom loss of cases. Analysis of losses and follow-up study may be used to reduce bias from sample mortality.

(*Go to* 7)

---

7 *Sources of bias from the investigator or research situation,* other than errors in sampling, primarily affect internal validity. Some errors may be design errors or errors in carrying out a procedure, such as in a drug study where one group may not comply with the treatment, or where the follow-up to treatment procedures varies among groups. Natural setting program studies often suffer internal validity problems when the program that is implemented is in fact different from the program the investigator thought was being implemented. Five categories of bias from the investigator or the situation are discussed next.

(*Go to* 8)

---

8 A. *Experimenter effects* include a variety of effects related to the investigator's characteristics or behavior. Black interviewers get different responses than do white interviewers in some studies. Age or sex of the interviewer may make a difference. Interview responses may suffer from leading questions. Bias in favor of desired results often occurs. Blind and double-blind designs, where possible, help to guard against experimenter effects. Such experimenter bias, sometimes called the Rosenthal effect, has occurred in classroom studies (one such set has been called the Pygmalion in the classroom studies) where students were randomly assigned expected performance levels and these levels were communicated to the teachers. The students, indeed, met these expected levels (sort of a self-fulfilling prophecy) under the influence of teacher expectation. Experimenter effects are a source of internal invalidity because the source of outcomes cannot be isolated.

(*Go to* 9)

---

9 B. *History* (the effect of nonexperimental events) is bias due to events occurring in the environment during the same time as the treatment under investigation. Such events are a potential source of internal invalidity; that is, such nonexperimental events may be a rival hypothesis for explaining results. Control groups are a principle means of dealing with history as a source of internal invalidity: if control and experimental groups have the same history, then difference can be attributed to something other than nonexperimental events.

C. *Maturation*, or change in the respondents or units being studied as a function of time not specific to particular events, is a source of internal invalidity. Such processes as growing older, hungrier, more tired, cell decay, group members getting to know one another, etc., may explain outcomes as well as the experimental treatment. Control groups, again, which will mature or develop in a similar way to the treatment groups are a primary means of controlling maturation effects.

(*Go to* 10)

---

10 D. *Instrumentation* variation is a source of internal invalidity. Observers or scorers may change their standards. Even machines, such as blood analyzers, will drift in their comparisons or measurement. Historical records may suffer from different recorders using different standards or redefining terms. A study may compare results of a skilled surgeon or test administrator against an inexperienced surgeon or test administrator.

E. *Statistical regression* is the tendency for cases with extreme scores to move toward the center scores (regress) on subsequent measurements. Any score in which there is natural variation within a subject over time (e.g., opinions), or in which there is good chance for measurement error or influence by extraneous factors will, by chance, show extreme cases in a distribution. On remeasurement, however, some other cases will, by chance, show the extreme values. Thus a before and after measurement using the high and low cases only will show movement of scores just by chance even without any treatment effect. Regression, as a source of internal invalidity, may be guarded against by using multiple selection criteria so as to increase the confidence of the initial selection, by using control groups to compare for nonexperimental changes, or by not focusing just on extreme cases for analysis.

(*Go to* 11)

---

11 Consider the following study and identify the various potential sources of invalidity.

A sociologist is studying changes in family interaction related to racial integration of schools. A sample of volunteer families is selected. A pretest on racial acceptance is given. Data are collected monthly over a 2-year period from observation in the home by the investigator, self-reports by the family members, and interviews with parents and their 5- to 7-year-old children. Riots occur in some schools during the study. Some families drop out of the study out of fear.

Given the list of potential problems below, identify which best illustrates history as a source of internal invalidity.

A. Volunteer sample is not representative of all school families.
B. Pretests and observations will sensitize the sample.
C. Self-reports will be selectively biased.
D. Changes in the families, particularly in 5- to 7-year-olds, will occur over the 2-year study.
E. School riots will produce effects on behavior in addition to school integration.
F. Sample mortality from drop-outs will change sample results.

When you have made your selection . . .

(*Go to* 12)

---

12 History, or the effects of nonexperimental events as a potential source of invalidity, is best illustrated by E, the school riots. Without further design control, any changes in measures of racial acceptance or family interaction may be explained by the impact of school riots as well as by school integration.

Now using the same study and list of potential problems, identify which problem best illustrates maturation as a source of internal invalidity.

When you have made your selection . . .

(*Go to* 13)

---

13 Maturation, or changes related to time, as a potential source of error for internal validity, is best illustrated by D, the changes in families and children over the 2-year time. Young children, we know in particular, will not be interacting with their families in the same way at age 5 as at age 7.

Next, consider four *sources of bias from the respondent* that include both internal and external validity problems.

(*Go to* 14)

---

14 A. *Role selection or test reaction* in which respondents alter their usual behavior is a source of external invalidity. When subjects are sensitized to a situation by a pretest or when they select a role in an experiment, there may be a true treatment effect, but the results may not reoccur in a natural setting where such respondent reaction is not triggered.

B. *Reactive effects of the situation* can occur when the respondents are aware of the experimental situation. Similar to test reaction, situation reaction leads to external invalidity where results may occur but cannot be generalized to the population. Such a "guinea pig effect" has been called the Hawthorne effect because of job production studies done years ago in the General Electric Hawthorne Works in Illinois. There, production increases occurred regardless of the experimental treatment due to the selected workers reacting to being singled out for study. Desensitization or treating both control and experimental group alike are ways to control situation reaction.

(*Go to* 15)

15 C. *Response set* is the tendency to behave in patterned ways that may be the explanation for an outcome rather than the treatment under investigation (a problem of internal validity). People are more likely to agree to statements than disagree. Acquiescence or yea-saying may produce apparent effects independent of the treatment. Sequencing of material combined with response set may bias responses to earlier or later material. As another example, studies measuring people movement must take into account a general right-turn response set of people on entering a building or room.

D. *Testing* or effects of measurement is a source of error from respondent reaction. A pretest or even the instructions may alter the subsequent responses to the treatment or posttest. A patient may be sensitized by an interview to give "acceptable" answers in a later interview. Such effects are a source of internal invalidity, masking the treatment effects. A control group without a pretest but with the experimental treatment is sometimes used to control for testing effects.

(*Go to* 16)

16 Consider the following study:

A school refers students with high aggressive behavior to Dr. Able. He selects the 25 in greatest need and works with them in a special program. Dr. Baker independently evaluates the students at the end of the program and finds them better adjusted and less aggressive

Continued.

than in Dr. Able's preprogram evaluation. The treatment program is judged successful.

Indicate which one or more of the following are serious threats to internal validity.

History ________	Instrumentation ________
Maturation ________	Regression ________
Testing ________	Sample mortality ________

When you have made your decision . . .

(*Go to* 17)

---

17 Serious threats to internal validity in the previous problem include history (major events in their lives may occur to the students during the course of treatment), maturation (self-control may be gained as they get older), instrumentation (differences in evaluation criteria between Dr. Able and Dr. Baker), and regression (selecting only the extreme cases).

The thrust of this text has been to prepare you for critical reading of the research literature in your field, with particular emphasis on statistics. In many cases you will have to depend on the scientific process of peer review and trust that the research you read was well done. But you also should be able to ask pertinent questions of a research report and expect the answers to be obtainable from the report. Such questions should include: Is the research question clear? Were the design and method sound? Are the findings clearly stated? Were the procedures understandable and well done? Was the statistical analysis appropriate? Were potential errors in the research and analysis well managed and explained? The following outline may be useful in evaluating research reports.

*Outline for evaluation of a research report**

1. Object or hypothesis
   a. What are the objectives of the study or the questions to be answered?
   b. What is the population to which the investigators intend to refer their findings?
2. Design of the investigation
   a. Was the study an experiment, investigation, survey, planned observation, or analysis of records?
   b. How was the sample selected? Are there possible sources of selec-

*Adapted from an outline by Colton, T. 1974. Statistics in medicine. Little, Brown & Co., Boston, by permission of the author and publishers.

tion that would make the sample atypical or nonrepresentative? If so, what provision was made to deal with this bias?

c. What is the nature of the control group or standard of comparison?

3. Observations
   a. Are there clear definitions of the terms used including treatment criteria, measurements made, and criteria of outcome?
   b. Was the method of classification or measurement consistent for all the subjects and relevant to the objectives of the investigation? Are there possible biases in measurement, and, if so, what provisions were made to deal with them?
   c. Are the observations reliable, valid, and reproducible?
4. Presentation of findings
   a. Are the findings presented clearly, objectively, and in sufficient detail to enable the reader to judge them for himself or herself?
   b. Are the findings internally consistent, i.e., do the numbers add up properly, can the different tables be reconciled, etc.?
5. Analysis
   a. Are the data worthy of statistical analysis? If so, are the methods of statistical analysis appropriate to the source and nature of the data, and is the analysis correctly performed and interpreted?
   b. Is there sufficient analysis to determine whether "significant differences" may in fact be due to lack of comparability of the groups in sex or age distributions or other relevant variables? Were potential sources of error in internal and external validity considered?
6. Conclusions
   a. Which conclusions are justified by the findings? Which are not?
   b. Are the conclusions relevant to the questions posed by the investigators?

(*Go to* 18)

---

18 Many other such questions have been raised in the preceding chapters. Statistics is a science that is one tool of the scientist. The logic of statistical inference and methodological control of sources of error are the fundamentals you should have obtained from studying this book. The fine points must come from further study or from working with those who specialize in statistics and research methodology.

(*End of chapter* 12)

---

appendix

A

Statistical tables

TABLE 1. Squares, square roots, and reciprocals

N	N^2	$\sqrt{N}$	$\sqrt{10N}$	$1/N$
1	1	1.0000	3.1623	1.000000
2	4	1.4142	4.4721	.500000
3	9	1.7321	5.4772	.333333
4	16	2.0000	6.3246	.250000
5	25	2.2361	7.0711	.200000
6	36	2.4495	7.7460	.166667
7	49	2.6458	8.3666	.142857
8	64	2.8284	8.9443	.125000
9	81	3.0000	9.4868	.111111
10	1 00	3.1623	10.0000	.100000
11	1 21	3.3166	10.4881	.090909
12	1 44	3.4641	10.9545	.083333
13	1 69	3.6056	11.4018	.076923
14	1 96	3.7417	11.8322	.071429
15	2 25	3.8730	12.2474	.066667
16	2 56	4.0000	12.6491	.062500
17	2 89	4.1231	13.0384	.058824
18	3 24	4.2426	13.4164	.055556
19	3 61	4.3589	13.7840	.052632
20	4 00	4.4721	14.1421	.050000
21	4 41	4.5826	14.4914	.047619
22	4 84	4.6904	14.8324	.045455
23	5 29	4.7958	15.1658	.043478
24	5 76	4.8990	15.4919	.041667
25	6 25	5.0000	15.8114	.040000
26	6 76	5.0990	16.1245	.038462
27	7 29	5.1962	16.4317	.037037
28	7 84	5.2915	16.7332	.035714
29	8 41	5.3852	17.0294	.034483
30	9 00	5.4772	17.3205	.033333

TABLE 1. Squares, square roots, and reciprocals—cont'd

N	N^2	$\sqrt{N}$	$\sqrt{10N}$	$1/N$
31	9 61	5.5678	17.6068	.032258
32	10 24	5.6569	17.8885	.031250
33	10 89	5.7446	18.1659	.030303
34	11 56	5.8310	18.4391	.029412
35	12 25	5.9161	18.7083	.028571
36	12 96	6.0000	18.9737	.027778
37	13 69	6.0828	19.2354	.027027
38	14 44	6.1644	19.4936	.026316
39	15 21	6.2450	19.7484	.025641
40	16 00	6.3246	20.0000	.025000
41	16 81	6.4031	20.2485	.024390
42	17 64	6.4807	20.4939	.023810
43	18 49	6.5574	20.7364	.023256
44	19 36	6.6332	20.9762	.022727
45	20 25	6.7082	21.2132	.022222
46	21 16	6.7823	21.4476	.021739
47	22 09	6.8557	21.6795	.021277
48	23 04	6.9282	21.9089	.020833
49	24 01	7.0000	22.1359	.020408
50	25 00	7.0711	22.3607	.020000
51	26 01	7.1414	22.5832	.019608
52	27 04	7.2111	22.8035	.019231
53	28 09	7.2801	23.0217	.018868
54	29 16	7.3485	23.2379	.018519
55	30 25	7.4162	23.4521	.018182
56	31 36	7.4833	23.6643	.017857
57	32 49	7.5498	23.8747	.017544
58	33 64	7.6158	24.0832	.017241
59	34 81	7.6811	24.2899	.016949
60	36 00	7.7460	24.4949	.016667
61	37 21	7.8102	24.6982	.016393
62	38 44	7.8740	24.8998	.016129
63	39 69	7.9373	25.0998	.015873
64	40 96	8.0000	25.2982	.015625
65	42 25	8.0623	25.4951	.015385
66	43 56	8.1240	25.6905	.015152
67	44 89	8.1854	25.8844	.014925
68	46 24	8.2462	26.0768	.014706
69	47 61	8.3066	26.2679	.014493
70	49 00	8.3666	26.4575	.014286
71	50 41	8.4261	26.6458	.014085
72	51 84	8.4853	26.8328	.013889
73	53 29	8.5440	27.0185	.013699
74	54 76	8.6023	27.2029	.013514
75	56 25	8.6603	27.3861	.013333
76	57 76	8.7178	27.5681	.013158
77	59 29	8.7750	27.7489	.012987

Continued.

TABLE 1. Squares, square roots, and reciprocals—cont'd

N	N^2	$\sqrt{N}$	$\sqrt{10N}$	$1/N$
78	60 84	8.8318	27.9285	.012821
79	62 41	8.8882	28.1069	.012658
80	64 00	8.9443	28.2843	.012500
81	65 61	9.0000	28.4605	.012346
82	67 24	9.0554	28.6356	.012195
83	68 89	9.1104	28.8097	.012048
84	70 56	9.1652	28.9828	.011905
85	72 25	9.2195	29.1548	.011765
86	73 96	9.2736	29.3258	.011628
87	75 69	9.3274	29.4958	.011494
88	77 44	9.3808	29.6648	.011364
89	79 21	9.4340	29.8329	.011236
90	81 00	9.4868	30.0000	.011111
91	82 81	9.5394	30.1662	.010989
92	84 64	9.5917	30.3315	.010870
93	86 49	9.6437	30.4959	.010753
94	88 36	9.6954	30.6594	.010638
95	90 25	9.7468	30.8221	.010526
96	92 16	9.7980	30.9839	.010417
97	94 09	9.8489	31.1448	.010309
98	96 04	9.8995	31.3050	.010204
99	98 01	9.9499	31.4643	.010101
100	1 00 00	10.0000	31.6228	.010000

TABLE 2. Z scores and proportions of area under the normal curve

The area between the mean (Z = 0) and Z standard deviation units away from the mean (column B), and the area beyond Z (column C), expressed as a proportion of the total area under the curve.

This table shows:

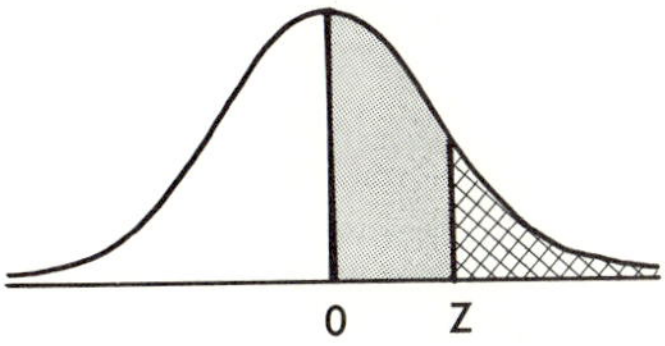

(A) Z	(B) *Area between mean and Z*	(C) *Area beyond Z*	(A) Z	(B) *Area between mean and Z*	(C) *Area beyond Z*
0.00	.0000	.5000	0.25	.0987	.4013
0.01	.0040	.4960	0.26	.1026	.3974
0.02	.0080	.4920	0.27	.1064	.3936
0.03	.0120	.4880	0.28	.1103	.3897
0.04	.0160	.4840	0.29	.1141	.3859
0.05	.0199	.4801	0.30	.1179	.3821
0.06	.0239	.4761	0.31	.1217	.3783
0.07	.0279	.4721	0.32	.1255	.3745
0.08	.0319	.4681	0.33	.1293	.3707
0.09	.0359	.4641	0.34	.1331	.3669
0.10	.0398	.4602	0.35	.1368	.3632
0.11	.0438	.4562	0.36	.1406	.3594
0.12	.0478	.4522	0.37	.1443	.3557
0.13	.0517	.4483	0.38	.1480	.3520
0.14	.0557	.4443	0.39	.1517	.3483
0.15	.0596	.4404	0.40	.1554	.3446
0.16	.0636	.4364	0.41	.1591	.3409
0.17	.0675	.4325	0.42	.1628	.3372
0.18	.0714	.4286	0.43	.1664	.3336
0.19	.0753	.4247	0.44	.1700	.3300
0.20	.0793	.4207	0.45	.1736	.3264
0.21	.0832	.4168	0.46	.1772	.3228
0.22	.0871	.4129	0.47	.1808	.3192
0.23	.0910	.4090	0.48	.1844	.3156
0.24	.0948	.4052	0.49	.1879	.3121

Portions of this table were taken from Croxton, F. E. 1949. Tables of areas in two tails and in one tail of the normal curve. Copyright by Prentice-Hall, Inc., Englewood Cliffs, N. J.

Continued.

TABLE 2. Z scores and proportions of area under the normal curve—cont'd

(A) Z	(B) *Area between mean and Z*	(C) *Area beyond Z*	(A) Z	(B) *Area between mean and Z*	(C) *Area beyond Z*
0.50	.1915	.3085	0.90	.3159	.1841
0.51	.1950	.3050	0.91	.3186	.1814
0.52	.1985	.3015	0.92	.3212	.1788
0.53	.2019	.2981	0.93	.3238	.1762
0.54	.2054	.2946	0.94	.3264	.1736
0.55	.2088	.2912	0.95	.3289	.1711
0.56	.2123	.2877	0.96	.3315	.1685
0.57	.2157	.2843	0.97	.3340	.1660
0.58	.2190	.2810	0.98	.3365	.1635
0.59	.2224	.2776	0.99	.3389	.1611
0.60	.2257	.2743	1.00	.3413	.1587
0.61	.2291	.2709	1.01	.3438	.1562
0.62	.2324	.2676	1.02	.3461	.1539
0.63	.2357	.2643	1.03	.3485	.1515
0.64	.2389	.2611	1.04	.3508	.1492
0.65	.2422	.2578	1.05	.3531	.1469
0.66	.2454	.2546	1.06	.3554	.1446
0.67	.2486	.2514	1.07	.3577	.1423
0.68	.2517	.2483	1.08	.3599	.1401
0.69	.2549	.2451	1.09	.3621	.1379
0.70	.2580	.2420	1.10	.3643	.1357
0.71	.2611	.2389	1.11	.3665	.1335
0.72	.2642	.2358	1.12	.3686	.1314
0.73	.2673	.2327	1.13	.3708	.1292
0.74	.2704	.2296	1.14	.3729	.1271
0.75	.2734	.2266	1.15	.3749	.1251
0.76	.2764	.2236	1.16	.3770	.1230
0.77	.2794	.2206	1.17	.3790	.1210
0.78	.2823	.2177	1.18	.3810	.1190
0.79	.2852	.2148	1.19	.3830	.1170
0.80	.2881	.2119	1.20	.3849	.1151
0.81	.2910	.2090	1.21	.3869	.1131
0.82	.2939	.2061	1.22	.3888	.1112
0.83	.2967	.2033	1.23	.3907	.1093
0.84	.2995	.2005	1.24	.3925	.1075
0.85	.3023	.1977	1.25	.3944	.1056
0.86	.3051	.1949	1.26	.3962	.1038
0.87	.3078	.1922	1.27	.3980	.1020
0.88	.3106	.1894	1.28	.3997	.1003
0.89	.3133	.1867	1.29	.4015	.0985

TABLE 2. Z scores and proportions of area under the normal curve—cont'd

(A) Z	(B) *Area between mean and Z*	(C) *Area beyond Z*	(A) Z	(B) *Area between mean and Z*	(C) *Area beyond Z*
1.30	.4032	.0968	1.70	.4554	.0446
1.31	.4049	.0951	1.71	.4564	.0436
1.32	.4066	.0934	1.72	.4573	.0427
1.33	.4082	.0918	1.73	.4582	.0418
1.34	.4099	.0901	1.74	.4591	.0409
1.35	.4115	.0885	1.75	.4599	.0401
1.36	.4131	.0869	1.76	.4608	.0392
1.37	.4147	.0853	1.77	.4616	.0384
1.38	.4162	.0838	1.78	.4625	.0375
1.39	.4177	.0823	1.79	.4633	.0367
1.40	.4192	.0808	1.80	.4641	.0359
1.41	.4207	.0793	1.81	.4649	.0351
1.42	.4222	.0778	1.82	.4656	.0344
1.43	.4236	.0764	1.83	.4664	.0336
1.44	.4251	.0749	1.84	.4671	.0329
1.45	.4265	.0735	1.85	.4678	.0322
1.46	.4279	.0721	1.86	.4686	.0314
1.47	.4292	.0708	1.87	.4693	.0307
1.48	.4306	.0694	1.88	.4699	.0301
1.49	.4319	.0681	1.89	.4706	.0294
1.50	.4332	.0668	1.90	.4713	.0287
1.51	.4345	.0655	1.91	.4719	.0281
1.52	.4357	.0643	1.92	.4726	.0274
1.53	.4370	.0630	1.93	.4732	.0268
1.54	.4382	.0618	1.94	.4738	.0262
1.55	.4394	.0606	1.95	.4744	.0256
1.56	.4406	.0594	1.96	.4750	.0250
1.57	.4418	.0582	1.97	.4756	.0244
1.58	.4429	.0571	1.98	.4761	.0239
1.59	.4441	.0559	1.99	.4767	.0233
1.60	.4452	.0548	2.00	.4772	.0228
1.61	.4463	.0537	2.01	.4778	.0222
1.62	.4474	.0526	2.02	.4783	.0217
1.63	.4484	.0516	2.03	.4788	.0212
1.64	.4495	.0505	2.04	.4793	.0207
1.65	.4505	.0495	2.05	.4798	.0202
1.66	.4515	.0485	2.06	.4803	.0197
1.67	.4525	.0475	2.07	.4808	.0192
1.68	.4535	.0465	2.08	.4812	.0188
1.69	.4545	.0455	2.09	.4817	.0183

Continued.

TABLE 2. Z scores and proportions of area under the normal curve—cont'd

(A) Z	(B) *Area between mean and Z*	(C) *Area beyond Z*	(A) Z	(B) *Area between mean and Z*	(C) *Area beyond Z*
2.10	.4821	.0179	2.50	.4938	.0062
2.11	.4826	.0174	2.51	.4940	.0060
2.12	.4830	.0170	2.52	.4941	.0059
2.13	.4834	.0166	2.53	.4943	.0057
2.14	.4838	.0162	2.54	.4945	.0055
2.15	.4842	.0158	2.55	.4946	.0054
2.16	.4846	.0154	2.56	.4948	.0052
2.17	.4850	.0150	2.57	.4949	.0051
2.18	.4854	.0146	2.58	.4951	.0049
2.19	.4857	.0143	2.59	.4952	.0048
2.20	.4861	.0139	2.60	.4953	.0047
2.21	.4864	.0136	2.61	.4955	.0045
2.22	.4868	.0132	2.62	.4956	.0044
2.23	.4871	.0129	2.63	.4957	.0043
2.24	.4875	.0125	2.64	.4959	.0041
2.25	.4878	.0122	2.65	.4960	.0040
2.26	.4881	.0119	2.66	.4961	.0039
2.27	.4884	.0116	2.67	.4962	.0038
2.28	.4887	.0113	2.68	.4963	.0037
2.29	.4890	.0110	2.69	.4964	.0036
2.30	.4893	.0107	2.70	.4965	.0035
2.31	.4896	.0104	2.71	.4966	.0034
2.32	.4898	.0102	2.72	.4967	.0033
2.33	.4901	.0099	2.73	.4968	.0032
2.34	.4904	.0096	2.74	.4969	.0031
2.35	.4906	.0094	2.75	.4970	.0030
2.36	.4909	.0091	2.76	.4971	.0029
2.37	.4911	.0089	2.77	.4972	.0028
2.38	.4913	.0087	2.78	.4973	.0027
2.39	.4916	.0084	2.79	.4974	.0026
2.40	.4918	.0082	2.80	.4974	.0026
2.41	.4920	.0080	2.81	.4975	.0025
2.42	.4922	.0078	2.82	.4976	.0024
2.43	.4925	.0075	2.83	.4977	.0023
2.44	.4927	.0073	2.84	.4977	.0023
2.45	.4929	.0071	2.85	.4978	.0022
2.46	.4931	.0069	2.86	.4979	.0021
2.47	.4932	.0068	2.87	.4979	.0021
2.48	.4934	.0066	2.88	.4980	.0020
2.49	.4936	.0064	2.89	.4981	.0019

TABLE 2. Z scores and proportions of area under the normal curve—cont'd

(A) Z	(B) *Area between mean and Z*	(C) *Area beyond Z*
2.90	.4981	.0019
2.91	.4982	.0018
2.92	.4982	.0018
2.93	.4983	.0017
2.94	.4984	.0016
2.95	.4984	.0016
2.96	.4985	.0015
2.97	.4985	.0015
2.98	.4986	.0014
2.99	.4986	.0014
3.00	.4987	.0013
3.01	.4987	.0013
3.02	.4987	.0013
3.03	.4988	.0012
3.04	.4988	.0012
3.05	.4989	.0011
3.06	.4989	.0011
3.07	.4989	.0011
3.08	.4990	.0010
3.09	.4990	.0010
3.10	.4990	.0010
3.11	.4991	.0009
3.12	.4991	.0009
3.13	.4991	.0009
3.14	.4992	.0008
3.15	.4992	.0008
3.16	.4992	.0008
3.17	.4992	.0008
3.18	.4993	.0007
3.19	.4993	.0007
3.20	.4993	.0007
3.21	.4993	.0007
3.22	.4994	.0006
3.23	.4994	.0006
3.24	.4994	.0006
3.25	.4994	.0006
3.30	.4995	.0005
3.35	.4996	.0004
3.40	.4997	.0003
3.45	.4997	.0003
3.50	.4998	.0002
3.60	.4998	.0002
3.70	.4999	.0001
3.80	.4999	.0001
3.90	.49995	.00005
4.00	.49997	.00003

TABLE 3. Distribution of t

The 1-tailed level shows:

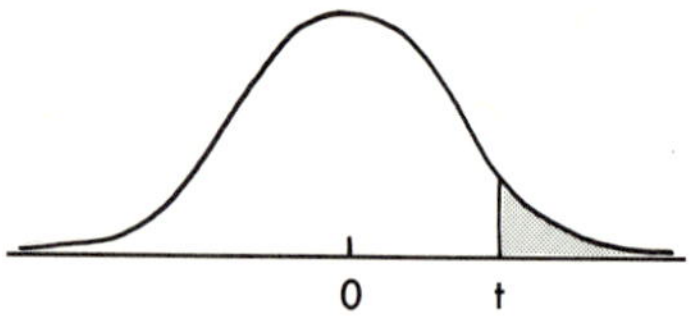

df	.10	.05	.025	.005	.0005
	Level of significance for 1-tailed test				
	Level of significance for 2-tailed test				
	.20	.10	.05	.01	.001
1	3.078	6.314	12.706	63.657	636.619
2	1.886	2.920	4.303	9.925	31.598
3	1.638	2.353	3.182	5.841	12.941
4	1.533	2.132	2.776	4.604	8.610
5	1.476	2.015	2.571	4.032	6.859
6	1.440	1.943	2.447	3.707	5.959
7	1.415	1.895	2.365	3.499	5.405
8	1.397	1.860	2.306	3.355	5.041
9	1.383	1.833	2.262	3.250	4.781
10	1.372	1.812	2.228	3.169	4.587
11	1.363	1.796	2.201	3.106	4.437
12	1.356	1.782	2.179	3.055	4.318
13	1.350	1.771	2.160	3.012	4.221
14	1.345	1.761	2.145	2.977	4.140
15	1.341	1.753	2.131	2.947	4.073
16	1.337	1.746	2.120	2.921	4.015
17	1.333	1.740	2.110	2.898	3.965
18	1.330	1.734	2.101	2.878	3.922
19	1.328	1.729	2.093	2.861	3.883
20	1.325	1.725	2.086	2.845	3.850

TABLE 3. Distribution of t—cont'd

	Level of significance for 1-tailed test				
	.10	.05	.025	.005	.0005
	Level of significance for 2-tailed test				
df	.20	.10	.05	.01	.001
21	1.323	1.721	2.080	2.831	3.819
22	1.321	1.717	2.074	2.819	3.792
23	1.319	1.714	2.069	2.807	3.767
24	1.318	1.711	2.064	2.797	3.745
25	1.316	1.708	2.060	2.787	3.725
26	1.315	1.706	2.056	2.779	3.707
27	1.314	1.703	2.052	2.771	3.690
28	1.313	1.701	2.048	2.763	3.674
29	1.311	1.699	2.045	2.756	3.659
30	1.310	1.697	2.042	2.750	3.646
40	1.303	1.684	2.021	2.704	3.551
60	1.296	1.671	2.000	2.660	3.460
120	1.289	1.658	1.980	2.617	3.373
∞	1.282	1.645	1.960	2.576	3.291

Portions of this table were taken from Table III of Fisher, R. A., and F. Yates. 1963. Statistical tables for biological, agricultural and medical research, ed. 6. Longman Group Ltd., London (previously published by Oliver & Boyd Ltd., Edinburgh). Reprinted by permission of the authors and publishers.

TABLE 4. Distribution of χ^2

This table shows:

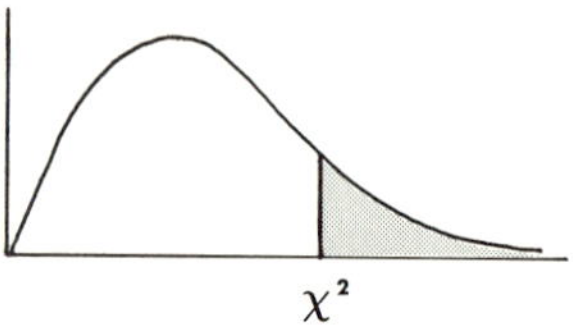

df	Level of significance .10	.05	.02	.01	.001
1	2.706	3.841	5.412	6.635	10.827
2	4.605	5.991	7.824	9.210	13.815
3	6.251	7.815	9.837	11.341	16.268
4	7.779	9.488	11.668	13.277	18.465
5	9.236	11.070	13.388	15.086	20.517
6	10.645	12.592	15.033	16.812	22.457
7	12.017	14.067	16.622	18.475	24.322
8	13.362	15.507	18.168	20.090	26.125
9	14.684	16.919	19.679	21.666	27.877
10	15.987	18.307	21.161	23.209	29.588
11	17.275	19.675	22.618	24.725	31.264
12	18.549	21.026	24.054	26.217	32.909
13	19.812	22.362	25.472	27.688	34.528
14	21.064	23.685	26.873	29.141	36.123
15	22.307	24.996	28.259	30.578	37.697

TABLE 4. Distribution of χ^2—cont'd

df	Level of significance .10	.05	.02	.01	.001
16	23.542	26.296	29.633	32.000	39.252
17	24.769	27.587	30.995	33.409	40.790
18	25.989	28.869	32.346	34.805	42.312
19	27.204	30.144	33.687	36.191	43.820
20	28.412	31.410	35.020	37.566	45.315
21	29.615	32.671	36.343	38.932	46.797
22	30.813	33.924	37.659	40.289	48.268
23	32.007	35.172	38.968	41.638	49.728
24	33.196	36.415	40.270	42.980	51.179
25	34.382	37.652	41.566	44.314	52.620
26	35.563	38.885	42.856	45.642	54.052
27	36.741	40.113	44.140	46.963	55.476
28	37.916	41.337	45.419	48.278	56.893
29	39.087	42.557	46.693	49.588	58.302
30	40.256	43.773	47.962	50.892	59.703

Portions of this table were taken from Table IV of Fisher, R. A., and F. Yates. 1963. Statistical tables for biological, agricultural and medical research, ed. 6. Longman Group Ltd., London (previously published by Oliver & Boyd Ltd., Edinburgh). Reprinted by permission of the authors and publishers.

TABLE 5. Values of r

Critical values of the product-moment correlation coefficient for testing H_0: r = 0
The 1-tailed level shows:

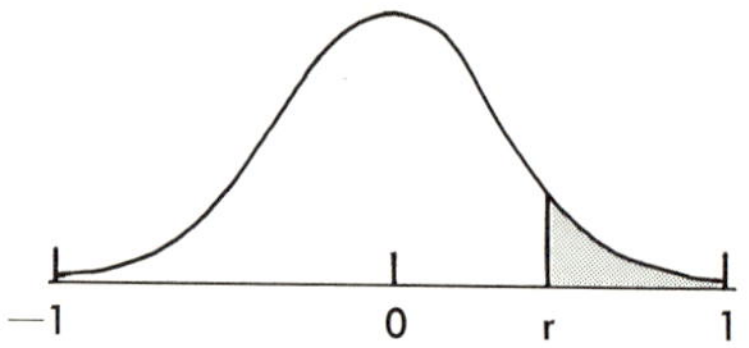

df	Level of significance for 1-tailed test			
	.05	.025	.005	.0005
	Level of significance for 2-tailed test			
(N − 2)	.10	.05	.01	.001
5	.669	.754	.874	.951
6	.621	.707	.834	.925
7	.582	.666	.798	.898
8	.549	.632	.765	.872
9	.521	.602	.735	.847
10	.497	.576	.708	.823
11	.476	.553	.684	.801
12	.458	.532	.661	.780
13	.441	.514	.641	.760
14	.426	.497	.623	.742
15	.412	.482	.606	.725
16	.400	.468	.590	.708
17	.389	.456	.575	.693
18	.378	.444	.561	.679
19	.369	.433	.549	.665
20	.360	.423	.537	.652
22	.344	.404	.515	.629
24	.330	.388	.496	.607
26	.317	.374	.479	.588
28	.306	.361	.463	.570

TABLE 5. Values of r—cont'd

df (N − 2)	Level of significance for 1-tailed test: .05	.025	.005	.0005
	Level of significance for 2-tailed test: .10	.05	.01	.001
30	.296	.349	.449	.554
35	.275	.325	.418	.519
40	.257	.304	.393	.490
45	.243	.288	.372	.465
50	.231	.273	.354	.443
60	.211	.250	.325	.408
70	.195	.232	.302	.380
80	.183	.217	.283	.357
90	.173	.205	.267	.338
100	.164	.195	.254	.321
200	.116	.138	.181	.23
300	.095	.113	.148	.19
400	.082	.098	.128	.16
500	.074	.088	.115	.15
1000	.052	.062	.081	.10

Portions of this table were taken from Table VII of Fisher, R. A., and F. Yates. 1963. Statistical tables for biological, agricultural and medical research, ed. 6. Longman Group Ltd., London (previously published by Oliver & Boyd Ltd., Edinburgh). Reprinted by permission of the authors and publishers.

TABLE 6. Transformations for r to Z and Z to r

r	Z	*r*	Z	*r*	Z
.01	.010	.34	.354	.67	.811
.02	.020	.35	.365	.68	.829
.03	.030	.36	.377	.69	.848
.04	.040	.37	.388	.70	.867
.05	.050	.38	.400	.71	.887
.06	.060	.39	.412	.72	.908
.07	.070	.40	.424	.73	.929
.08	.080	.41	.436	.74	.950
.09	.090	.42	.448	.75	.973
.10	.100	.43	.460	.76	.996
.11	.110	.44	.472	.77	1.020
.12	.121	.45	.485	.78	1.045
.13	.131	.46	.497	.79	1.071
.14	.141	.47	.510	.80	1.099
.15	.151	.48	.523	.81	1.127
.16	.161	.49	.536	.82	1.157
.17	.172	.50	.549	.83	1.188
.18	.182	.51	.563	.84	1.221
.19	.192	.52	.576	.85	1.256
.20	.203	.53	.590	.86	1.293
.21	.214	.54	.604	.87	1.333
.22	.224	.55	.618	.88	1.376
.23	.234	.56	.633	.89	1.422
.24	.245	.57	.648	.90	1.472
.25	.256	.58	.663	.91	1.528
.26	.266	.59	.678	.92	1.589
.27	.277	.60	.693	.93	1.658
.28	.288	.61	.709	.94	1.738
.29	.299	.62	.725	.95	1.832
.30	.309	.63	.741	.96	1.946
.31	.321	.64	.758	.97	2.092
.32	.332	.65	.775	.98	2.298
.33	.343	.66	.793	.99	2.647

TABLE 7. Critical values of the rank order correlation coefficient for testing H_0: $\rho = 0$

	Level of significance for 1-tailed test			
	.05	.025	.005	.0005
	Level of significance for 2-tailed test			
N	.10	.05	.01	.001
5	.900	1.000	—	—
6	.829	.886	1.000	—
7	.714	.786	.929	1.000
8	.643	.738	.881	.976
9	.600	.700	.833	.933
10	.564	.648	.794	.903
12	.503	.587	.727	.846
14	.464	.538	.679	.802
16	.429	.503	.635	.762
18	.401	.472	.600	.728
20	.380	.447	.570	.696
22	.361	.425	.544	.667
24	.344	.406	.521	.642
26	.331	.390	.501	.619
28	.317	.375	.483	.598
30	.306	.362	.467	.580
40	.264	.313	.405	.507
60	.214	.255	.331	.418
100	.165	.197	.257	.326

From Zar, J. 1972. Significance testing of the Spearman rank correlation coefficient. J. Am. Stat. Assoc. **67:**578-580. Reprinted by permission of the author and publisher.

appendix
B

Answers to examinations

Each answer is referenced to a specific chapter and objective. If you missed the answer, you might want to refer to that chapter and objective and restudy the content material.

Chapters 1 to 5

Question	*Answer*	*Chapter*	*Objective*
1. A.	A	1	1
B.	B	1	1
2. A.	2	1	2
B.	4	1	2
C.	5	1	2
D.	1	1	2
E.	3	1	2
3.	Nominal, ordinal, interval, ratio	1	4
4. A.	Ordinal	1	7
B.	Interval or ratio	1	7
C.	Interval or ratio	1	7
D.	Ordinal	1	7
E.	Nominal	1	7
F.	Nominal	1	7

Question	*Answer*	*Chapter*	*Objective*
5. A.	18- to 39-year-olds	2	6
B.	Headache	2	2
C.	Less; 1.6:1 vs. 1.875:1	2	1, 2, 6
6.	Group 1 changed the most (.625 vs. −.20). Group 1's hostility index increased, but group 2's index decreased after the experiment.	2	4, 5
7. A.	18-29 or 12	3 5	1 3
B. C. D.	(see table below)	3	2, 3, 4

Interval	*f*	*%*	*Cum. f*	*Cum. %*
28-29	1	3.3	30	100.0
26-27	2	6.7	29	96.7
24-25	1	3.3	27	90.0
22-23	2	6.7	26	86.7
20-21	16	53.3	24	80.0
18-19	8	26.7	8	26.7
Total	30	100.0		

Question	*Answer*	*Chapter*	*Objective*
E.	Right; most scores are on low end with tail to right (high, positive) end.	3	7
8.	Refer to Fig. 3-9 for answer.	3	8
9. A.	Dispersion	5	1
B.	Central tendency	4	1
10. A.	4	4	2
B.	1	4	2
C.	2	5	2
D.	3	4	2
E.	5	5	2
11. A.	20	4	3
B.	20	4	3
C.	20.87	4	3
D.	Mean, given age as interval data	4	5

Question	*Answer*	*Chapter*	*Objective*
E.	Mode requires nominal; median requires ordinal; mean requires interval.	4	5
F.	Mean is pulled in direction of skew; i.e., toward high end in the example.	4	9
G.	Mean is probably best because there is not much difference between three measures; i.e., skew is not severe. Severe skew would suggest use of median.	4	8
12. A.	$P_{75} = Q_3 = 21.32$; or 75% of all the ages of juniors lie below 21.32 years.	5	4, 6
B.	$P_{25} = Q_1 = 19.38$; 25% of all the ages of juniors lie below 19.38 years.	5	4, 6
C.	$Q = \frac{21.32 - 19.38}{2} = \frac{1.94}{2} = 0.97$ years	5	5
D.	Middle 50% of ages varies less than 1 year (0.97), but full range of ages varies 12 years. Q is more stable between samples than is range.	5	6, 7
13. A.	68.26% or approximately 2/3	5	9
B.	95.44% or approximately 95%	5	9
C.	$s_{\text{Group I}} > s_{\text{Group II}}$ ($s_I = 5$; $s_{II} = 3$)	5	8
14.	$s = 2.57$	5	10
15.	The mean and median are the same for each group and approximately the same between groups, with the women being a bit more positive toward ERA than the male sample. However, the female sample feels more consistently about the ERA than does the male sample. Approximately 66% of women rate the ERA between 4.0 and 8.2, whereas the same proportion of males rate the ERA between 2.5 and 9.3.	5	11
16. A.	1	5	13
B.	4	5	13
C.	2	5	13
D.	8	5	13

Question	Answer	Chapter	Objective
E.	7	5	13
F.	5	5	13
G.	6	5	13
H.	3	5	13

Chapters 6 to 11

Question	Answer	Chapter	Objective
1. A.	1	6	2
B.	0	6	2
2. A.	$\frac{1}{3} + \frac{1}{3} = \frac{2}{3}$	6	3
B.	$\frac{10}{30} \times \frac{10}{29} = \frac{100}{870} = \frac{10}{87}$	6	6
3.	A and C	6	7
4.	C	7	3
5.	B	7	3
6. A.	Internal	7	3
B.	Internal	7	3
C.	External	7	3
D.	Internal	7	3
7.	Induction	7	2
8. A.	Disaster training	7	7
B.	Learning or experience during disaster	7	7
C.	Age and education	7	7
D.	Rating of favorability	7	7
9.	Deduction	7	2

Question		*Answer*	*Chapter*	*Objective*
10.		A	7	3
11.		C	7	4
12.	A.	Reliability	7	8
	B.	Validity	7	8
13.		1 or 2, since: $P(0) = \frac{32}{243}$ $P(1) = \frac{80}{243}$ $P(2) = \frac{80}{243}$ $P(3) = \frac{40}{243}$ $P(4) = \frac{10}{243}$ $P(5) = \frac{1}{243}$	8	1
14.	A.	Sampling	8	4
	B.	Standard error of the statistic	8	4
15.	A.	Yes	9	2
	B.	No	9	2
	C.	No	9	2
	D.	Yes	9	2
16.	A.	Mean IQ scores of poverty level children in Gary equal the mean IQ scores of poverty level children in Kansas City.	9	3
	B.	(1) The proportion of blacks who are in-migrants to Kansas City is greater than (lesser than) that of Gary. (2) The proportion of blacks who are in-migrants to Kansas City is not the same as that of Gary.	9	3
17.	A.	Type I (α) error is the probability of rejecting a hypothesis when in fact it is true.	9	5
	B.	Type II (β) error is the probability of not rejecting a hypothesis when in fact it is false.	9	6

<table>
<tr><th>Question</th><th></th><th>Answer</th><th>Chapter</th><th>Objective</th></tr>
<tr><td>18.</td><td></td><td>Decreases</td><td>9</td><td>7</td></tr>
<tr><td>19.</td><td></td><td>$p = .01$ (or 1%)</td><td>9</td><td>8</td></tr>
<tr><td>20.</td><td></td><td>B</td><td>10</td><td>3</td></tr>
<tr><td>21.</td><td>A.</td><td>3</td><td>10</td><td>1, 4</td></tr>
<tr><td></td><td>B.</td><td>6</td><td>10</td><td>2, 5</td></tr>
<tr><td></td><td>C.</td><td>7</td><td>10</td><td>10</td></tr>
<tr><td></td><td>D.</td><td>2</td><td>10</td><td>1, 4</td></tr>
<tr><td></td><td>E.</td><td>4</td><td>10</td><td>2, 5</td></tr>
<tr><td></td><td>F.</td><td>5</td><td>10</td><td>2, 5</td></tr>
<tr><td></td><td>G.</td><td>1</td><td>10</td><td>1,4</td></tr>
<tr><td>22.</td><td>A.</td><td>There is evidence that the sample clinics are not performing at the level anticipated. ($t = -3.265$, 1 tail, $df = 24$, Table $t = -1.711$; reject at $p < .05$.)</td><td>10</td><td>7, 9</td></tr>
<tr><td></td><td>B.</td><td>Low and high dogmatism groups differ significantly in their attitudes toward federally funded poverty programs; low dogmatism persons are more in favor. ($t = 2.025$, 1 tail, $df = 18$, Table $t = 1.734$; $p < .05$.)</td><td>10</td><td>7, 9</td></tr>
<tr><td></td><td>C.</td><td>
<table>
<tr><th>Social class aspirations</th><th>Sex
Male</th><th>Female</th><th></th></tr>
<tr><td>High</td><td>6</td><td>15</td><td>21</td></tr>
<tr><td>Medium</td><td>12</td><td>7</td><td>19</td></tr>
<tr><td>Low</td><td>7</td><td>3</td><td>10</td></tr>
<tr><td></td><td>25</td><td>25</td><td>50</td></tr>
</table>
There is evidence to indicate that class aspirations are not independent of sex; girls tended to aspire higher than boys. ($\chi^2 = 6.774$, $df = 2$, Table $\chi^2 = 5.991$; $p < .05$.)</td><td>10</td><td>10, 12, 13, 14, 15</td></tr>
<tr><td></td><td>D.</td><td>The salesman's claim is not legitimate; we reject an H_0 of no difference between $p_s = .80$ and $p_u = .90$. ($Z = 4.714$, 1 tail, Table $Z = 1.64$; $p < .05$.)</td><td>10</td><td>7, 9</td></tr>
<tr><td></td><td>E.</td><td>There is evidence that supports the detail man's claim; the physician should seriously consider switching to the new drug, Xpheno. ($Z = -2.169$, 1 tail, Table $Z = -1.64$; $p < .05$.)</td><td>10</td><td>7, 9</td></tr>
</table>

Question	*Answer*	*Chapter*	*Objective*
F.	The examination results suggest using the self-learning approach since the H_0 of no difference between the two methodologies was rejected. ($Z = -1.98$, 2 tail, Table $Z = -1.96$; $p < .05$.)	10	7, 9
G.	There is reason to suspect that fathers with retarded children have lower incomes than the state incomes as a whole. ($Z = -2.704$, 1 tail, Table $Z = -1.64$; $p < .05$.)	10	7, 9
23.	The 95% confidence interval for the mean of this sample runs from $7524.71 to $8097.29.	10	19, 20
24. A.	Correlation	11	1
B.	Prediction	11	1
25. A.	Appropriate, ρ	11	2, 3
B.	Inappropriate	11	2, 3
C.	Appropriate, r	11	2, 3
D.	Inappropriate	11	2, 3
26.	Yes	11	14
27.	$-.6$	11	6
28. A.	$-.804$	11	8
B.	As income increases, weeks premature decreases; results are significant. ($r = -.804$, 1 tail, df = 8, Table $r = .549$; $p < .05$.)	11	9
C.	.34 to .95	11	10
D.	$Y = 12 - 1.5X$	11	15
E.	$Y = 7.5$ weeks premature	11	12
29. A.	$\rho = -.875$	11	16, 17
B.	For a 1-tailed test $p < .025$; the relationship between headache frequency and class rank is statistically significant; the lower the class rank, the greater the frequency of headaches.	11	18

appendix C

Formulas

Following are the formulas used throughout this text. They are keyed to the chapter in which they are first presented and are meant as a reference for the learner.

Chapter 2. The basic comparison of numbers: ratios

1. Proportion: some part of the total divided by the total, presented in decimal form; e.g., 25/100 = 0.25
2. Percentage: proportion multiplied by 100; e.g., 0.25 × 100 = 25%
3. Rate of change $= \dfrac{\text{Measurement now} - \text{Measurement then}}{\text{Measurement then}}$

Chapter 3. Frequency distributions and graphics

4. Range $= (X_H - X_L) + 1$
   where X_H = raw score of highest value
   X_L = raw score of lowest value
5. Cumulative proportion: divide cumulative frequency of each interval by total number of scores

$$\text{Cumulative proportion} = \frac{\text{Cumulative frequency of interval X}}{\text{Total number}}$$

Chapter 4. Measures of central tendency

6. Mean $= \overline{X} = \dfrac{X_1 + X_2 + X_3 + \ldots + X_n}{N} = \dfrac{\Sigma X}{N}$
7. Mean of means: To calculate the mean of means use a weighted mean, i.e., the number of scores used for each mean, times that mean is a weighted mean:

$$\overline{\overline{X}} = \frac{\Sigma f\overline{X}}{\Sigma f}$$

Chapter 5. Measures of dispersion

8. Percentiles (also quartiles and median):

$$P_x = \frac{\text{Lower real limit of interval}}{\text{containing the wanted number}} + \left(\frac{\text{Number of added cases}}{\text{Frequency (number of cases in interval)}} \times \frac{\text{Width of}}{\text{interval}} \right)$$

9. Semi-interquartile range:

$$Q = \frac{Q_3 - Q_1}{2}$$

10. Standard deviation:

$$s = \sqrt{\frac{\Sigma(X - \overline{X})^2}{N}} \quad \text{or} \quad s = \sqrt{\frac{\Sigma X^2}{N} - \overline{X}^2}$$

Chapter 6. Basic probability

11. Probability for equally likely events $= \dfrac{\text{Number of events of interest}}{\text{Total possible events}}$
12. $p(A \text{ or } B) = p(A) + p(B)$ when A and B are mutually exclusive
13. $p(A \text{ and } B) = p(A) \times p(B)$ when A and B are independent
14. The number of combinations of N things r at a time $= \binom{N}{r} = \dfrac{N!}{r!\,(N - r)!}$
15. The probability of r given N in a binomial $= \binom{N}{r} p^r q^{N-r}$
16. Standard error of the mean $= \sigma_{\overline{x}} = \dfrac{\sigma}{\sqrt{N}}$

Chapter 10. Tests of statistical significance

17. Critical value or significance test $= \dfrac{\text{Obtained statistic} - \text{Expected statistic}}{\text{Standard error of obtained statistic}}$
18. General confidence interval (large N) = Obtained statistic $\pm (Z_\alpha)$(Standard error of obtained statistic)
19. General confidence interval (small N) = Obtained statistic $\pm (t_\alpha)$(Standard error of obtained statistic)
20. Mean, 1 sample, large N:

$$Z = \frac{\overline{X} - \mu}{\sigma_{\overline{x}}} = \frac{\overline{X} - \mu}{\sigma_x / \sqrt{N}}$$

21. Proportion, 1 sample, large N:

$$Z = \frac{p - p_u}{\sigma_p} = \frac{p - p_u}{\sqrt{p_u q_u / N}}$$

22. Mean, 1 sample, small N:

$$t = \frac{\overline{X} - \mu}{\hat{\sigma}_{\overline{x}}} = \frac{\overline{X} - \mu}{s/\sqrt{N - 1}} \qquad df = N - 1$$

23. Means, 2 independent samples, large N:

$$Z = \frac{(\overline{X}_1 - \overline{X}_2) - (\mu_1 - \mu_2)}{\sigma_{\overline{x}_1 - \overline{x}_2}} = \frac{\overline{X}_1 - \overline{X}_2}{\sqrt{\dfrac{s_1^{\,2}}{N_1} + \dfrac{s_2^{\,2}}{N_2}}}$$

24. Proportions, 2 independent samples, large N:

$$Z = \frac{(p_1 - p_2) - (p_{u_1} - p_{u_2})}{\hat{\sigma}_{p_1 - p_2}} = \frac{p_1 - p_2}{\sqrt{\hat{p}_u\hat{q}_u}\sqrt{\frac{N_1 + N_2}{N_1N_2}}}$$

$$\text{where } \hat{p}_u = \frac{N_1p_1 + N_2p_2}{N_1 + N_2} \text{ and } \hat{q}_u = 1 - \hat{p}_u$$

25. Means, 2 independent samples, small N:

$$t = \frac{(\overline{X}_1 - \overline{X}_2) - (\mu_1 - \mu_2)}{\hat{\sigma}_{\overline{X}_1 - \overline{X}_2}} = \frac{\overline{X}_1 - \overline{X}_2}{\sqrt{\frac{N_1s_1^2 + N_2s_2^2}{N_1 + N_2 - 2}}\sqrt{\frac{N_1 + N_2}{N_1N_2}}}$$

$$df = N_1 + N_2 - 2$$

26. $\chi^2 = \Sigma\frac{(f_o - f_e)^2}{f_e} = \Sigma\frac{f_o^2}{f_e} - N$

For contingency table case, df = (number of rows − 1)(number of columns − 1)

Chapter 11. Association

27. $r = \frac{\Sigma(X - \overline{X})(Y - \overline{Y})}{Ns_Xs_Y} = \frac{N\Sigma XY - \Sigma X\Sigma Y}{\sqrt{[N\Sigma X^2 - (\Sigma X)^2][N\Sigma Y^2 - (\Sigma Y)^2]}}$

$df = N - 2$

28. Confidence interval for r:

$$Z \pm Z_\alpha \frac{1}{\sqrt{N - 3}}$$

where Z = the Z transformation for r

29. General equation for a straight line: $Y = a + bX$

30. $b = \frac{\Sigma(X - \overline{X})(Y - \overline{Y})}{\Sigma(X - \overline{X})^2} = \frac{N\Sigma XY - \Sigma X\Sigma Y}{N\Sigma X^2 - (\Sigma X)^2}$

31. $a = \overline{Y} - b\overline{X} = \frac{\Sigma Y}{N} - b\frac{\Sigma X}{N}$

32. $\rho = 1 - \frac{6\Sigma D^2}{N(N^2 - 1)}$

References

Anders, T. R. 1971. Retrospective reports of retrieval from short-term memory. J. Exp. Psych. **90:**251-257.

Booth, A. 1972. Sex and social participation. Am. Sociol. Rev. **37:**183-192.

Born, D. G., and E. W. Herbert. 1971. A further study of personalized instruction for students in large university classes. J. Exp. Educ. **40:**6-11 (Fall).

Campbell, D. T., and J. C. Stanley. 1963. Experimental and quasi-experimental designs for research. Chicago: Rand McNally.

Colton, T. 1974. Statistics in medicine. Boston: Little, Brown & Co.

de Charms, R., and V. Carpenter. 1968. Measuring motivation in culturally disadvantaged school children. J. Exp. Educ. **37:**31-41 (Fall).

Duchastel, P. C., and P. F. Merrill. 1973. The effects of behavioral objectives on learning: a review of empirical studies. Rev. Educ. Res. **43:**53-69 (Winter).

Duff, R. S., and A. G. M. Campbell. 1973. Moral and ethical dilemmas in the special-care nursery. New Engl. J. Med. **289:**890-894.

Festinger, L. 1957. The theory of cognitive dissonance. Harper & Row, New York.

Greeley, A. M.· 1973. The "religious factor" and academic careers: another communication. Am. J. Sociol. **78:**1247-1255.

Hack, M., M. Brish, D. M. Serr, V. Insler, and B. Lunenfeld. 1970. Outcome of pregnancy after induced ovulation. J.A.M.A. **211:**794-796.

Hermalin, A. I., and R. Farley. 1973. The potential for residential integration in cities and suburbs: implications for the busing controversy. Am. Sociol. Rev. **38:**595-610.

Hong, L. K. 1972. A comparative analysis of extended kin visitations, cohabitations and anomie in rural and urban Hong Kong. Sociol. Soc. Res. **57:**43-54.

Hurst, C. E. 1972. Race, class, and consciousness. Am. Sociol. Rev. **37:**658-670.

Kasl, S. V., G. W. Brooks, and W. L. Rodgers. 1970. Serum uric acid and cholesterol in achievement behavior and motivation. J.A.M.A. **214:**1158-1164.

Kelman, H. C., and L. H. Lawrence. 1972. Violent man: American response to the trial of Lt. William L. Calley. Psychol. Today **6:**41-45 (June).

Laube, J. 1973. Psychological reactions of nurses in disaster. Nurs. Res. **22:**343-346.

McLaughlin, M., A. L. Linch, and R. D. Snee. 1973. Longitudinal studies of lead levels in a US population. Arch. Environ. Health **27:**305-311.

Millimet, C. R., and D. F. Gardner. 1972. Induction of threat to self-esteem and the arousal and resolution of affect. J. Exp. Soc. Psychol. **8:**467-481.

Sargent, F. II. 1955. An application of statistics. Science **121:**402.

Sethi, B. B., S. S. Nathawat, and S. C. Gupta. 1973. Depression in India. J. Soc. Psychol. **91:**3-13.

Skipper, J. K., A. L. Guenther, and G. Nass. 1967. The sacredness of .05: a note concerning the uses of statistical levels of significance in social science. Am. Sociol. **2:**16-18 (Feb.).

Stahl, S. M. 1971. Illness among the aged: a study of the determinants of the perception of levels of health in an indigent, urban population. Unpublished Ph.D. dissertation. Urbana, Ill.: University of Illinois, Department of Sociology.

Stahl, S. M., and J. D. Hennes. 1973. Biostatistics: an experiment with self-learning in the health sciences. J. Med. Educ. **48:**271-275.

Stein, Z., and M. Susser. 1970. Mutability of intelligence and epidemiology of mild mental retardation. Rev. Educ. Res. **40:**29-67 (Feb.).

Toomey, M. 1972. Conflict theory approach to decision making applied to alcoholics. J. Personal. Soc. Psychol. **24:**199-206 (Nov.).

Tutt, N. S. 1973. Achievement motivation and delinquency. Br. J. Soc. Clin. Psychol. **12:**225-230.

Twain, M. 1883. Life on the Mississippi. Boston: J. B. Osgood.

Whitman, J. R. 1969. Free recall of verbal material in the classroom as a function of method of presentation and method of practice. J. Exp. Educ. **37:**78-81 (Summer).

Williams, J. A., Jr., N. Babchuk, and D. R. Johnson. 1973. Voluntary associations and minority status: a comparative analysis of Anglo, black, and Mexican Americans. Am. Sociol. Rev. **38:**637-646.

Zimet, S. G., J. L. Wiberg, and G. E. Blom. 1971. Attitudes and values in primers from the United States and twelve other countries. J. Soc. Psychol. **84:**167-170.

Index